D1161049

Elements of Ceramics
SECOND EDITION

F. H. NORTON

Professor Emeritus,
Massachusetts Institute of Technology

ADDISON-WESLEY PUBLISHING COMPANY

Reading, Massachusetts
Menlo Park, California · London · Amsterdam · Don Mills, Ontario · Sydney

This book is in the
ADDISON-WESLEY SERIES IN METALLURGY AND MATERIALS

Consulting Editor
Morris Cohen

ISBN 0-201-05306-3
CDEFGHIJK-HA-798

Preface

In the twenty-year period since the first edition of this book was published, there have been many changes in all branches of the ceramic field. Notable ones are float glass, man-made diamonds, glass ceramics, ferroelectric and ferromagnetic bodies, isostatic pressing, and computer batching. During this period the production of ceramic ware in the United States has tripled.

It was therefore decided that this book should be brought up to date and cover as thoroughly as space permitted new products and processes as well as the traditional ones.

The glaze and body formulas, small kiln designs, and other information intended for the artist have been omitted since this material is taken up more extensively in the author's *Ceramics for the Artist Potter*.

References to pertinent literature are placed at the end of each chapter. Unfortunately it has been found necessary to limit them to those in the English language as few of our undergraduate students have a reading knowledge in any others.

The author wishes to acknowledge the assistance given him by many in the ceramic field: Dr. Emil Deeg and Dr. Walter Sigmund provided help in the field of glass, Dr. William Prindle and Professor Arthur Friedberg in enamels, Mr. W. H. Wendel in abrasives, and Dr. Stanley Warshaw in whitewares. Dr. Morris Cohen was most helpful in assisting with the lay-out and scope of this book.

Many others have contributed illustrations which are acknowledged individually.

F.H.N.
August 1973

Contents

Chapter 1   **Introduction**

1. Importance of ceramics. *1* 2. Lines of ceramic development. *1*
3. Magnitude of the ceramic industry in the United States. *4*
4. Literature in the ceramic field. *4*

Chapter 2   **Crystal Physics**
1. Introduction. *8* 2. Elements of crystal physics. *8* 3. Silicate
structures. *13* 4. Kaolin minerals. *15* 5. Three-layer minerals. *18*
6. Hydrated aluminous minerals. *20* 7. Spinel structure. *22*
8. Preovskite structure. *22*

Chapter 3   **Clay**
1. Introduction. *24* 2. Kaolin. *24* 3. Ball clays. *26* 4. Fire clays. *26*
5. Properties of raw clays. *28* 6. Plastic properties. *36* 7. Dried
properties. *36* 8. Slaking properties. *36* 9. Firing properties. *36*

Chapter 4   **Feldspars and Other Fluxes**

1. Introduction *38* 2. Feldspar Minerals. *38* 3. Occurrence of feldspars. *39* 4. Mining and milling feldspar. *39* 5. Composition of
feldspars. *40* 6. Other fluxing materials. *40*

Chapter 5   **Other Minerals**

1. Introduction. *42* 2. Silica. *42* 3. Magnesite. *46* 4. Lime ($CaCO_3$) *47*
5. Dolomite. *47* 6. Chromite. *47* 7. Silicate minerals. *48* 8. Aluminous
minerals. *50* 9. Pure oxides. *50* 10. Carbonaceous materials. *50*
11. Relative abundance of the elements. *53*

Chapter 6   Mining and Treatment of the Raw Materials

1. Introduction *55* 2. Mining. *55* 3. Methods of comminution. *57* 4. Size classification. *60* 5. Disintegration. *64* 6. Chemical treatments. *64* 7. Magnetic separation. *66* 8. Froth flotation. *66* 9. Filtering. *66* 10. Drying. *66* 11. Storage and handling. *68*

Chapter 7   Particulate Solids and Water

1. Introduction. *72* 2. Elements of colloidal chemistry. *72* 3. Deflocculation of slips. *73* 4. Laws of flow. *76* 5. Mechanism of plasticity. *84* 6. Measurement of flow. *88* 7. Systems with nonclay materials. *89*

Chapter 8   Forming Methods

1. Introduction. *92* 2. Body preparation. *92* 3. Slip casting. *95* 4. Forming from soft plastic masses. *100* 5. Stiff plastic forming. *104* 6. Dry pressing. *105* 7. Dust pressing electronic ceramics. *109* 8. Isostatic pressing. *109* 9. Green finishing. *110*

Chapter 9   Drying Ceramic Ware

1. Introduction. *114* 2. Internal flow of moisture. *114* 3. Surface evaporation. *116* 4. Drying shrinkage. *117* 5. Achievement of maximun drying rate. *120* 6. Green strength. *122* 7. Types of dryer. *122*

Chapter 10   Firing Ceramic Ware

1. Introduction. *126* 2. Thermodynamics of reactions. *126* 3. Phase rule. *128* 4. Equilibrium diagrams. *128* 5. Reaction rates. *132* 6. Solid-state reactions. *132* 7. Means for measuring thermochemical changes. *134* 8. Thermochemical reactions in clays and other materials. *136* 9. Effect of heat on ceramic bodies. *140* 10. Solid-state sintering. *143* 11. Hot pressing pure oxides. *145* 12. Setting methods. *145* 13. Finishing fired ware. *151*

Chapter 11   Kilns

1. Introduction. *154* 2. Periodic kilns. *154* 3. Continuous kilns. *157* 4. Kiln efficiency. *159*

Chapter 12   The Glassy State

1. Introduction. *163* 2. Constitution of glass. *163* 3. Elastic and viscous forces in the glass network. *167* 4. Devitrification. *171*

Chapter 13   Glass Melting

1. Introduction. *173* 2. Glass compositions. *173* 3. Mechanism of melting. *175* 4. Melting equipment. *178*

Chapter 14   Glass Forming

1. Introduction. *184* 2. Forming methods. *184* 3. Finishing and annealing. *191* 4. Grinding and polishing. *191*

Chapter 15   Glazes

1. Introduction. *193* 2. Methods of expressing glaze compositions. *193* 3. Methods of compounding glazes. *196* 4. Application of glazes. *196* 5. Firing the glaze. *201* 6. Fitting the glaze to the body. *202* 7. Some examples of glazes. *206* 8. Glaze defects. *209*

Chapter 16   Enamels on Metal

1. Introduction. *211* 2. Low-fusing glasses. *211* 3. Theories of adherence. *212* 4. Methods of obtaining opacity. *212* 5. Jewelry enamels. *213* 6. Enamels for sheet steel. *214* 7. Cast-iron enamels. *219* 8. Enamels for aluminum. *222* 9. Refractory enamels. *222* 10. Development of new enamels. *222*

Chapter 17   Decorative Processes

1. Introduction. *223* 2. Modeling. *223* 3. Printing methods. *225* 4. Photographic method. *226* 5. Other processes. *226* 6. Mechanism of color formation in glasses. *228* 7. Solution colors. *231* 8. Colloidal colors. *233* 9. Colors in crystals. *235* 10. Ceramic stains. *235* 11. Lusters. *237* 12. Gilding. *237*

Chapter 18   Fine Ceramics

1. Introduction. *239* 2. Expressing body compositions. *239* 3. Triaxial bodies. *241* 4. Bodies for electrical and electronic uses. *243* 5. Other bodies. *245*

Chapter 19   Refractories and Insulators

1. Introduction. *252* 2. Heavy refractories. *252* 3. Insulating firebrick. *254* 4. Pure oxide refractories. *255* 5. Nonoxide refractory bodies. *256* 6. Refractory plastics, concretes, and mortars. *258* 7. Insulating materials. *258*

Chapter 20   Abrasives
1. Introduction. *260* 2. Natural abrasives. *260* 3. Man-made abrasives. *261* 4. Frinding wheels. *263* 5. Coated abrasives. *268* 6. Loose abrasives. *268*

Chapter 21   Ceramic Building Materials

1. Introduction. *269* 2. Building brick. *269* 3. Other clay products. *271* 4. Sand-lime brick. *271* 5. Lime. *271* 6. Portland cement. *272* 7. High-alumina cement. *275* 8. Gypsum plaster. *275* 9. Oxychloride cements. *277* 10. Silicate cements *277* 11. Phosphate cements. *271*

Chapter 22   **Glass Products**

1. Introduction. *279* 2. Sheet glass. *279* 3. Plate glass. *279* 4. Container glass. *279* 5. Tableware and kitchen glass. *280* 6. Art glass. *280* 7. Special glasses. *280* 8. Optical glass *280* 9. Fiber glass. *280*

Chapter 23   **Properties of Ceramic Materials**

1. Introduction. *283* 2. Mechanical properties. *283* 3. Thermal properties. *286* 4. Electrical properties. *290* 5. Optical properties. *291*

Appendix   Table A.1. Properties of the atom *293*
Table A.2. Equivalent weights of common ceramic materials *296*
Table A.3. Temperature conversion table *297*
Table A.4. Temperature equivalents of Orton pyrometric cones *298*

Author Index *299*

Subject Index *303*

1

Introduction

1. Importance of Ceramics

Ceramics is a field primarily concerned with the treatment of nonmetallic minerals by various processes, including heat, to produce articles with aesthetic or utilitarian properties. Looking about our homes, we can see ceramics on every side; glass in the windows, tile and plumbing fixtures in the bathroom, tableware in the dining room, brick in the fireplace, and cement in the cellar floor. Our automobiles are made largely of steel produced in refractory lined furnaces and precisely formed with abrasive wheels. Our eye-glasses are made from carefully controlled ophthalmic glass and our television sets are filled with hundreds of pieces of electronic ceramic materials. Thus it is quite evident that our life is dependent on the many products of the great and diversified ceramic industry.

2. Lines of Ceramic Development

Our present ceramic products did not, of course, emerge all at one time, but followed definite lines of development over many centuries. It is of interest to examine some of the more important periods of development to try and discover the driving force that pushed them forward.

One of the most prodigious expansions of ceramics occurred in China around the start of the Christian era. From a coarse earthenware there evolved during three or four centuries, first, strong watertight stoneware and, second, translucent porcelain of great beauty. What caused this rapid advance? There was, of course, the availability of partially decomposed feldspar as a natural body material and the evolvement of high-temperature kilns. But what seems to be the important factor was the intense desire of the Chinese people to produce and live with things of beauty. The potter was not hurried or pressed for quantity; only quality mattered. A subsidized potter did not think it unusual that he should spend his whole lifetime to produce just one exquisite vase for the Emperor's palace.

1

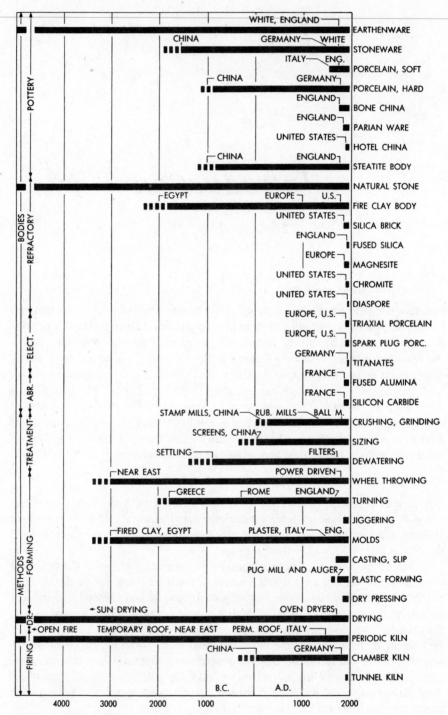

Fig. 1.1   Technical history of ceramics.

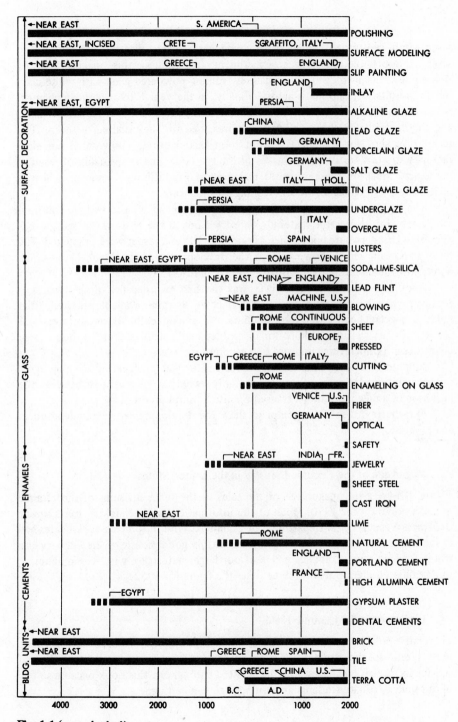

Fig. 1.1 (*concluded*)

A few pieces of this porcelain trickled over the trade routes into Europe in the seventeenth century, at which time crude earthenware was the rule. Potters in several European countries began experiments to reproduce this amazing ware. At first some translucent ware was made by mixing powdered glass with white clay, but the results were not satisfactory. Finally, in the year 1709, a German chemist, Graf von Tschirnhaus, discovered the secret of making a body of kaolin, feldspar, and quartz fired far above the temperatures used for the usual earthenware. The secret was kept for a short time, but soon leaked out to allow small porcelain factories to start up in many places in Europe under the sponsorship of wealthy noblemen. Most of these ventures had a short life, although some, like Sèvres, Meissen, and Copenhagen, are still turning out fine ware.

The eighteenth century in England was a period of intense experimentation, during which innumerable materials were tried out in the potteries of Wedgwood, Spode, and others where bodies such as Parian, basalt, and jasper were originated. The most important result, however, was the development of the unique bone-china body, now the mainstay of the British pottery industry.

Another period of great activity was the first two decades of the twentieth century, when large amounts of electrical power became available to vitalize the field of electrochemistry. Fused alumina and silicon carbide for abrasives and refractories, as well as graphite for electrodes and crucibles, were made in large quantities at temperatures unheard of a generation earlier.

Anyone following the history of scientific and engineering developments cannot help wondering why, almost invariably, twenty years must pass between the inception of a new idea and its translation into a useful product.

The charts in Fig. 1.1 attempt to show the development of ceramic products and processes.

3. Magnitude of the Ceramic Industry in the United States

Figure 1.2 gives the magnitude of the sales in the main divisions of the ceramic industry for the year 1970. Much of the information has been taken from *Current Industrial Reports* (U.S. Department of Commerce) but in some cases estimates had to be made. This diagram should give a reasonably good picture of the industry as a whole for that year. Ceramics is one of our larger industries, with sales amounting to ten billion dollars annually, three times the figure for 1952.

4. Literature in the Ceramic Field

There are books on ceramics available, but it is felt that those listed below will be most useful to the student. It is regretted that the many excellent books and articles in foreign languages must be omitted throughout this book because so few of our undergraduate students are equipped to handle them.

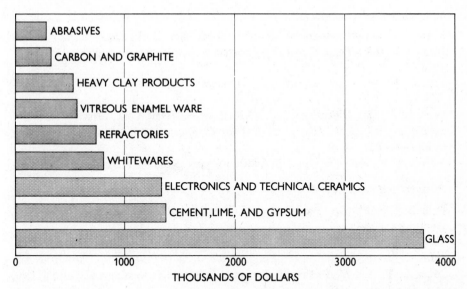

Fig. 1. 2   Magnitude of the ceramics industry in the United States for 1970.

History

Barber, E. A., *Pottery and Porcelain of the United States*, Putnam, New York, 1893
Solon, L. M., *The Art of the Old English Potter*, John Francis Co., New York, 1906
Cox, W. E., *Pottery and Porcelain*, 2 vols., Crown, New York, 1944

Raw materials

Reis, H., *Clays, Their Occurrence, Properties, and Uses*, 3rd ed., Wiley, New York, 1927
Mudd, S. W. (ed.), *Industrial Minerals and Rocks*, 2nd ed., Am. Inst. of Mining and Met. Engs., New York, 1949
Ladoo, R. B., and W. M. Myers, *Nonmetallic Minerals*, 2nd ed., McGraw-Hill, New York, 1951
Minerals Yearbook, U.S. Bureau of Mines, Government Printing Office, Washington, D.C., 1964
Ryan, W., *Properties of Ceramic Raw Materials*, North Staffordshire College of Technology, England, 1968

Calculations

Andrews, A. I., *Ceramic Tests and Calculations*, Wiley, New York, 1928
Griffiths, R., and C. Radford, *Calculations in Ceramics*, Maclaren, London, 1965

Mineralogy and ceramic science

Kingery, W. D., *Introduction to Ceramics*, Wiley, New York, 1960
Burk, J. E. (ed.), *Progress in Ceramic Sciences*, 4 vols., Pergamon Press, New York, 1961–4
Fulrath, R. M., and J. A. Pask (eds.), *Ceramic Microstructures*, Wiley, New York, 1968
Grim, R. F., *Clay Mineralogy*, 2nd ed., McGraw-Hill, New York, 1968
Levine, E. M., *et al.*, *Phase Diagrams for Ceramists*, American Ceramic Society, Columbus, Ohio, 1969
Warren, B. E., *X-Ray Diffraction*, Addison-Wesley, Reading, Mass., 1969

Products and processes

Budnikov, P. P., *The Technology of Ceramics and Refractories*, M. I. T. Press, Cambridge, 1964
Van Vlack, L. H., *Physical Ceramics for Engineers*, Addison-Wesley, Reading, Mass., 1964
Jones, J. T., and M. F. Berard, Ceramics, Industrial Processing and Testing, Iowa State Uni. Press, Ames, Iowa, 1972
Census of Manufacturers, Government Printing Office, Washington, D.C., 1968
Promisel, N. E. (ed.), *Ceramic Processing*, Publication no. 1576, National Academy of Sciences, Washing, D.C., 1968
Norton, F. H., *Refractories*, McGraw-Hill, New York, 1968
Rouch, H. W., Sr., *et al.*, *Ceramic Fibers and Fibrous Composite Materials*, Academic Press, New York, 1968
Shaw, K., *Ceramic Colors and Pottery Decoration*, Praeger, New York, 1968
Stewart, G. H. (ed.), *Science of Ceramics*, British Ceramic Society, Stoke-on-Trent, England, 1970
Whittemore, O. J., Jr., *Electron Ceramics*, Special Publication no. 3, American Ceramic Society, Columbus, Ohio, 1970
Norton, F. H., *Fine Ceramics*, McGraw-Hill, New York, 1970
Manning, J. H., and A. Spence, *Ceramics for Engineers*, Morgan-Grampian, London, 1969
Singer, F., and S. S. Singer, *Industrial Ceramics*, Chemical Publication Co., New York, 1964

Glass

Volf, M. B., *Technical Glasses*, Pitman, London, 1961
Weyl, W. A., and C. E. Marboe, *The Constitution of Glasses*, Interscience, New York, 1962

Glossaries

Van Schoik, E. C., *Ceramic Glossary*, American Ceramic Society, Columbus, Ohio, 1963
Dodd, A. E., *Dictionary of Ceramics*, George Newnes, London, 1964

Periodicals

American Ceramic Society, Journal, Bulletin, and Abstracts, Columbus, Ohio
American Concrete Institute Journal, New York
Brick and Clay Record, Chicago
British Ceramic Society, Transactions, Journal, and Abstracts, Stoke-on-Trent, England
Ceramic Age, Cleveland, Ohio
Ceramic Industry, Chicago
Glass Industry, New York
Glass Technology, Sheffield, England

2

Crystal physics

1. Introduction

As the materials used in ceramics are largely crystalline it is desirable for those studying them to have a background in crystal structure. The way atoms are placed in the crystal lattice in an ordered array is a fascinating study in special packing. The subject is covered briefly in this chapter.

2. Elements of Crystal Physics

Properties of the atoms. The atom may be considered to consist of a nucleus having a net positive charge equal to the atomic number, surrounded by shells of electrons which total up to an equal negative charge. For example, hydrogen has one electron and an atomic number of one; electrons are added, one at a time, to fill shell after shell until Lawrencium, with a total of 103 electrons, is reached. Table 2.1 shows the electrons in each shell for the various atom species, and it will be seen later that the properties of the element are influenced by the number and location of the electrons.

When atoms or ions (an ion is an atom that has lost or gained one or more electrons) are packed together to form a crystal, each takes up a definite space that may be assumed to be a sphere. The radius of this sphere is known as the ionic radius, values of which are shown in Table A.1, of the Appendix. In a few cases, such as in the lead ions, the electrons are not symmetrically arranged about the nucleus, but are inclined to be lopsided or polarized. This tends to give a different type of packing than that indicated by the value of ionic radius alone, especially on the surface of solids.

The very simple model of the atom described above is adequate for a preliminary study of crystals and glasses. The real structure of the atom is more complex. There are a number of excellent treatises on the structure of the atom if the reader desires to go deeper into this subject.

Table 2.1 Classification of atoms according to structure

Complete shells (K L M N O P)	0	1	2	3	4	5	6	7	9	10	11	12	13	14	15	16	17	18	19	20	21	22	23	24	25	26	27	28	29	30	31
(outer shell electrons)	2	2	2	2	2	2	2	2	2	2	2	2	2	2	2	2	2	2	2	2	2	2	2	2	2	2	2	2	2	2	2
(outer shell electrons)									2	2	2	2	2	2	2	2	2	2	2	2	2	2	2	2	2	2	2	2	2	2	2
2	He	H																													
2 8	Ne	Li	Be	B	C	N	O	F																							
2 8 8	A	Na	Mg	Al	Si	P	S	Cl																							
2 8 8	Kr	K	Ca						Sc	Ti	V	Cr*	Mn	Fe	Co	Ni	Cu*														
2 8 18			Zn	Ga	Ge	As	Se	Br																							
2 8 18 8	Xe	Rb	Sr						Y	Zr	Nb*	Mo*	Ma*	Ru*	Rh*	Pd*	Ag*														
2 8 18 18			Cd	In	Sn	Sb	Te	I																							
2 8 18 18 8	Rn	Cs	Ba						Lu	Hf	Ta	W	Re	Os	Ir	Pt	Au*	La	Ce	Pr	Nd	Pm	Sm	Eu	Gd	Tb	Dy	Ho	Er	Tm	Yb
2 8 18 32 18			Hg	Tl	Pb	Bi	Po	At																							
2 8 18 32 18 8		Fr	Ra						Ac	Th*	Pa	U	Np	Pu	Am	Cm	Bk														

Annotations across the table:

- All shells complete / Few compounds / No absorption in visible spectrum *(column 0)*
- One shell incomplete / No absorption in visible spectrum
- Two shells incomplete / Variable valence / Broad band absorption in visible spectrum / (Transition elements)
- Three incomplete shells / Sharp band absorption in visible spectrum / (Rare earth elements)

*Elements in which the normal atom is believed to have one electron in the outer shell.

Bonding forces between atoms. When atoms are regularly located in the crystal lattice, forces must be present to hold them in place. These forces are termed *bonds* and generally represent a condition of balance between attraction forces and repulsion forces. Therefore, an additional force is required to move the atoms from their stable separation distance.

One of the important bonding forces in crystals is the ionic bond, which is due to the metallic atoms losing an outer electron to become positive ions and the nonmetallic atoms gaining an outer electron to form a negative ion. The resulting coulomb attraction holds the atoms together. Simple inorganic salts often have this type of bonding, for example, NaCl, as well as many of the minerals used in ceramics. Ionic bonded crystals are brittle and have medium to high melting points.

Another type of bonding force is the covalent bond, where a pair of electrons is shared by two atoms. Elements such as C, Si, N, P, and O often have the covalent bond. This type of bonding gives hard, strong materials with high melting points.

There is a third type of bond, called the metallic bond, in crystals composed only of positive ions. Here the closely packed atoms are pictured as surrounded by an electron cloud which holds them together. This bond gives more plastic materials with a wide range of melting points.

A fourth type of bond, which can hardly be called chemical bonding, is the van der Waals force. In general, it is weak in character and gives ready cleavage.

In any one crystal it cannot be assumed that the bonding force is exclusively one of those mentioned above; it is more generally a combination of them. This is certainly true in many of the silicates.

Unit cell. A crystal is made up of an orderly arrangement of one or more atomic species, as shown schematically by the two-dimensional network of Fig. 2.1(a). The

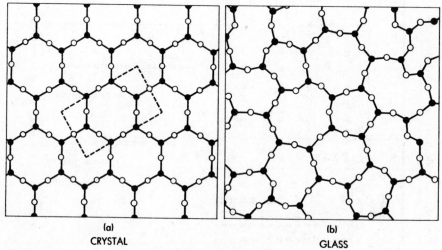

(a)
CRYSTAL

(b)
GLASS

Fig. 2. 1   Schematic structure of (a) crystal and (b) glass.

smallest portion of the crystal network that can be used as a repeating building block to form the whole is called a unit cell (outlined by dotted lines in the figure). The dimensions and the placing of the atoms in the unit cell are the fundamental property of the crystal. Most of the unit cells of the crystals used in ceramics have now been worked out on the basis of the pioneer work of W. L. Bragg and his students. Descriptions of these structures may be found in the *Zeitschrift für Kristallographie* and the references at the end of this chapter.

In contrast to the uniform pattern of the crystal is the random network of the glassy state as shown in Fig. 2.1(b). This state of solid matter will be discussed in Chapter 13. As an example of a crystal, the simple unit cell of CsCl type may be illustrated as in Fig. 2.2. Here the cell is cubic with the cations placed at the corners and the anion at the center. As only one-eighth of the cations are included at the corners, the cell represents $8 \times \frac{1}{8}$ cation $+ 1$ anion, which corresponds with the one-to-one formula of CsCl. The crystal is built up by repeating this cube over and over.

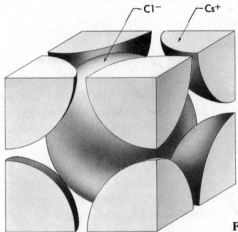

Fig. 2. 2   Unit cell of cesium chloride.

Anyone studying the atomic arrangement in crystals will be at once struck by the beauty of the patterns. Berkhoff in *Aesthetic Measure* has defined a beautiful design as one where the largest possible proportion of the elements involved shows a perceptible sense of order. The crystal network, therefore, is a three-dimensional design of great beauty. In fact, in most crystals with fixed relative numbers of the various atoms, there is only one possible arrangement that will give a stable structure of minimum energy. Each atom must fulfill certain rules in regard to bonding with other atoms. These rules were first set forth by Pauling, but little can be said about them here because of space limitations.

Coordination number. It has been shown that the atoms vary in diameter to a considerable extent. Thus it is possible to place a considerable number of small atoms about a large atom, while only a few large atoms can be placed about a small atom. Therefore, the ratio between the diameter of the cation and that of the anion is an important number, often referred to as the radius ratio. The number of atoms immediately surrounding and bonded to another atom is known as the coordination number. As the radius ratio increases, so does the coordination number, as shown in Table 2.2.

Table 2.2 Relations of radius ratio and coordination number for ions A and X acting as rigid spheres

Radius ratio $R_A:R_X$	Coordination number	Arrangement	Type of structure
	1		Single molecules
			Double molecules
Up to 0.15	2	Opposite each other	Molecular chains
0.15-0.22	3	Corners of an equi-lateral triangle	Boron nitride
0.22-0.41	4	Corners of a tetra-hedron	ZnS
0.41-0.73	6	Corners of an oc-tahedron	NaCl
0.73-1.00	8	Corners of a cube	CsCl
1.00 and above	12	Closest packing	Cu

Thus is can be seen that, for simple structures, where the ions act as rigid spheres, the type of packing in the crystal can be predicted with reasonable certainty. However, in cases where polarization occurs the ions are not spherical and the above rule requires some modification.

Crystal structure. The crystal is built up of a series of unit cells. A crystal starts growing from a nucleus of one or more unit cells forming a tiny group of atoms that may be thought of as the seed. This nucleus, under the proper environmental conditions, will grow by the addition of atoms according to a regular structural pattern. No crystal, however, is so perfect that all the atoms are in their proper places, for here and there discontinuities are present that cause defects in the structure and may have an important influence on the properties of the crystal.

The crystal has an outside form which permits it to be classified into six systems based on symmetry, but not all crystals are so perfect that the system can be determined by visual inspection. However, optical and x-ray studies will usually yield the symmetry relations. The six systems with their axes are shown in Fig. 2.3. There are many excellent treatises on crystallography, so that those interested may easily pursue this fascinating subject further.

Polymorphism. A specific compound such as silica (SiO_2) may occur in several different forms of crystal. In each form the ions and the ratio of the number of cations to anions is the same, but the arrangement differs. Some forms are stable in one temperature range and some in others. Even in small amounts, impurities may

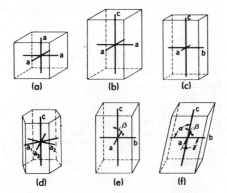

Fig. 2.3 The six crystallographic systems: (a) isometric; (b) tetragonal; (c) orthorhombic; (d) hexagonal; (e) monoclinic; (f) triclinic.

influence the occurrence of a particular form. These different crystalline forms of the same material, known as polymorphic forms, are of great interest in ceramics.

Solid solutions. In many cases it is possible to have a crystal A-X form a continuous series of mixtures with a crystal B-X. This may occur by B ions replacing some of the A ions (substitution solid solution); by A ions, if smaller than B, finding holes in B-X and gradually filling them (interstitial solid solution); or by omission of some ions from the lattice to form holes (omission solid solution). To form a solid solution the two crystals must be similar in geometry. Also, the ionic radius of A must not be more than 15% larger or smaller than that of B. The closer in size they are, the more complete will be the range of solid solution. Also, at higher temperatures, where the crystal lattice is loosened up by thermal agitation, solid solution can occur more readily.

3. Silicate Structures

At the present time the structures of the silicates are well known, based as they are on the early x-ray diffraction work of W. L. Bragg, B. E. Warren, and their followers.

The fundamental building block in the silicates is the silicon-oxygen tetrahedron with a silicon atom in the center. These tetrahedrons fit together in various ways to form silicates, as shown in Table 2.3. In this table no attempt is made to indicate the size of these atoms, only their centers. It should be realized that the large oxygen atoms determine the structure while the small silicon atoms fit into the holes.

The orthosilicates are of interest, for the independent tetrahedrons make a structure of good stability and thus form compounds with high fusion points. The disilicates have a sheet structure with perfect cleavage parallel to the *a-b* plane of the crystal. Mica and the clay minerals fall into this class.

Table 2.3 Structure of the silicates

	Motif of Si–O tetrahedrons	Si:O ratio of smallest unit	Volume charge of silica unit	Typical mineral	Atomic plan
Orthosilicates	Independent tetrahedrons sharing no oxygens	1:4	-4	Forsterite Mg_2SiO_4	
Pyrosilicates	Independent pairs of tetrahedrons sharing one oxygen	2:7	-6	Akermanite $Ca_2MgSi_2O_7$	
Metasilicate chains	Continuous single chains of tetrahedrons sharing two oxygens	1:3	-2	Diopside $CaMg(SiO_3)_2$	
Metasilicate chains	Continuous double chains of tetrahedrons sharing alternately 2 and 3 tetrahedrons	4:11	-6	Tremolite $H_2Ca_2Mg_5(SiO_3)_8$	
Metasilicate rings	Closed independent rings of tetrahedrons each sharing two oxygens	3:9 6:18	-6 -12	Benitoite $Ba\,Ti\,Si_3O_9$ Beryl $Al_2Be_3Si_6O_{18}$	
Disilicates	Continuous sheets of tetrahedrons each sharing three oxygens	4:10	-4	Muscovite $Al_4K_2(Si_6Al_2)O_{20}(OH)_4$	
Silica	Three-dimensional network of tetrahedrons each sharing all four oxygens	1:2 1:2	0 0	Quartz SiO_2 Orthoclase $K\,AlSi_3O_8$	(Three-dimensional structure not shown)

4. Kaolin Minerals

Formulas for the clay minerals. In the classic paper on *Minerals of the Montmorillonite Group* by Ross and Hendricks, a logical method of expressing the formula of a clay mineral from its chemical analysis is developed. As this method of writing the formulas will be used in this book, a brief discussion of the reasoning behind it will be necessary.

A general formula for montmorillonite, for example, is

$$\underbrace{[Al^{+3}_{a-y} + Fe^{+3}_{b} + Mg^{+2}_{d}\ ]}_{\substack{\text{Octahedral} \\ \text{coordination}}} \quad \underbrace{[Al^{+3}_{y} + Si_{4-y}]}_{\substack{\text{Tetrahedral} \\ \text{coordination}}}$$

$$\underbrace{O_{10}\,[OH]_2}_{\text{Anions}} \quad \underbrace{X_{0.33},}_{\substack{\text{Exchangeable} \\ \text{bases}}}$$

where the ions in octahedral positions $= a - y + b + d$, and 0.33 is the amount of the exchangeable bases.

From specific chemical analysis the values of the subscripts may be computed as shown in the paper cited above. For a typical montmorillonite, the expression becomes

$$\begin{bmatrix} Na_{0.33} \\ \uparrow \\ Al_{1.67}Mg_{0.33} \end{bmatrix} \quad [Si_4\,O_{10}]\,[OH]_2,$$

which is now completely balanced, the first bracket representing the cations in octahedral coordination.

Referring to Table A.1 in the Appendix, we see that the ionic radii of the cations and oxygen are as follows:

Ions	Mg^{++}	Al^{+++}	Si^{++++}	O^{--}
Ionic radius, Å	0.66	0.51	0.42	1.32
$\dfrac{\text{Cation radius}}{\text{O radius}}$	0.50	0.39	0.32	

By looking at Table 2.2, it will be seen that Mg and Al must be in sixfold (octahedral) coordination, while silicon must be fourfold (tetragonal) coordination.

Kaolinite $[(OH)_4\,Al_2\,Si_2\,O_5\,]$. The majority of high-grade clays consist largely of the mineral kaolinite. This mineral occurs in tiny flat plates roughly hexagonal in outline, as shown in Fig. 2.4 from an excellent photomicrograph taken with an electron microscope. The average size is about 0.7 micron in diameter and about 0.05 micron in thickness. At times these crystals are found in groups known as "books" or "worms", where many crystals are piled one on another.

Fig. 2.4   Kaolinite crystals with shadows three times the thickness, taken by C. E. Hall of M.I.T.

By means of optical and x-ray measurements, it has been possible to arrive at a probable arrangement of the atoms in the structure. As it is difficult to show this structure completely in a two-dimensional diagram, a dissected view is shown in Fig. 2.5.

The crystal is made up of a series of parallel sheets in the plan of the *a, b* axes, built up to the number of fifty in a crystal such as shown in Fig. 2.4. Each sheet is composed of a tetrahedral Si-O layer and an octahedral Al-OH layer, as shown in Fig. 2.5. While the sheet is unsymmetrical, the total charge adds up to zero, giving a balanced structure. The Si-O layer shows a close packing of the O^{--}, but with holes in the hexagonal structure. On the other hand, the Al-OH layer is close-packed with no holes.

In the diagram, the triclinic angles are not shown, as they are in planes normal to the paper, but they actually are 104.5 and 91.8° because of a shift of one layer over the next. The bonding between layers is due to weak, residual, or van der Waals forces between the O^{--} and OH^{-} ions, which accounts for the easy cleavage. It can be seen that the unit cell consists of one sheet containing four molecular weights of $(OH)_4 Al_2 Si_2 O_5$.

If possible, the student should study a three-dimensional model of this mineral, since it is difficult to portray this intricate structure clearly on the printed page. In fact, one of the best ways to become thoroughly familiar with the structure is to actually make models from appropriately sized spheres.

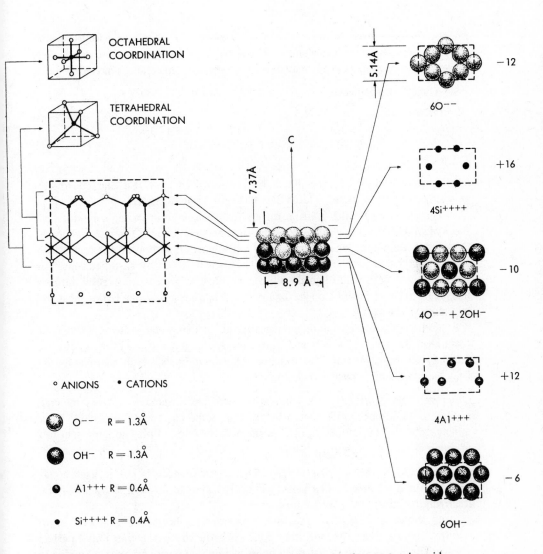

Fig 2.5 The unit cell of kaolinite, $2(OH)_4 Al_2 Si_2 O_5$. In the center is a side view of the unit cell. At the right is a series of horizontal sections through the cell at the height of the atomic layers. On the left is a schematic diagram of the side of the unit cell magnified two times to show the bonding.

There are several other minerals that are believed to have a sheet structure nearly identical to that of kaolinite; however, the shift from one sheet to another is in different directions in these minerals. They include dickite and nacrite, very similar to kaolinite, but rather rare constituents of clay, and halloysite, of finer grain size, less perfectly crystalline, and perhaps with a different arrangement of the sheets. Finally, there is the more or less amorphous allophane with very little regularity in the atomic arrangement. The latter two minerals are found in relatively pure form in few deposits but may become of some use in ceramics in the future.

5. Three-Layer Minerals

Montmorillonite

$$\begin{pmatrix} Na_{0.33} \\ \uparrow \\ Al_{1.67}Mg_{0.33} \end{pmatrix} Si_4 O_{10}(OH)_2.$$

This mineral is found in bentonite, derived from volcanic ash. It is characterized by very fine platelike particles, seldom over 0.05 micron in diameter.

The structure of this mineral differs from kaolinite in having a symmetrical sheet of an Al-OH octahedral layer sandwiched between two Si-O tetrahedral layers. One of the aluminum ions is replaced by a divalent ion such as magnesium, which gives a net charge for the unit cell of -1. This is balanced by adsorbed ions. The sheets are not definitely aligned with one another, so we cannot exactly define a unit cell or the monoclinic angle. The unit cell contains only one lattice sheet and two molecular weights.

Montmorillonite is unique among minerals in that water molecules can force themselves between the sheets and thus cause swelling. Also, ions may be adsorbed not only on the edges but also on the faces between sheets, which accounts for the large base exchange capacity of this mineral.

Other montmorillonite types of minerals. There are a number of other minerals similar to montmorillonite that are found in some clays and soils. While these minerals are of great interest to the soil chemist, they do not play an important role in ceramics.

Muscovite $[Al_4K_2(Si_6Al_2)O_{20}(OH)_4]$. This common mica is not in itself a clay mineral, but it is found as an accessory in many clays and forms an end member for the hydrated micas with claylike properties.

As shown in Fig. 2.6, muscovite consists of sheets similar to those of montmorillonite, but with potassium ions tying the sheets together into a perfect crystal. The K^+ ions are in the open holes of the Si-O layer and weakly bonded to 12 O^{--}'s. The substitution of an aluminum ion for one of silicon in the tetrahedral layer balances the charge of the potassium ion. It should be remembered that even though the aluminum ion is too large to go into tetrahedral coordination alone, it may substitute for silicon in the tetrahedral layer to a limited extent.

The bonding of the potassium ions is weak, which accounts for the perfect cleavage of this mineral, but the bonds are strong enough to cause each layer to be accurately aligned with the next and to prevent water molecules from getting between the layers. The monoclinic angle of 95° is a result of the shift in the octahedral layer, not of the shifting of one sheet over another. The unit cell consists of two sheets and contains two molecular weights.

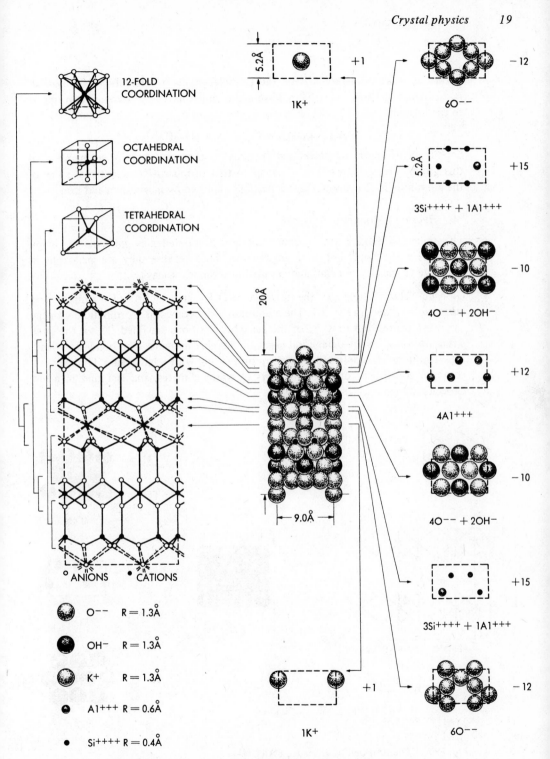

Fig. 2. 6 The unit cell of muscovite, $2Al_4K_2(Si_6Al_2)O_{20}(OH)_4$.

Micaceous clay minerals. These rather indeterminate minerals are important constituents of many clays. They have been named *illites* by Grim and correspond to the formula

$$Al_{4-a+b}Mg_aFe_6K_2Si_{8-y}Al_yO_{20}(OH)_4.$$

Sericite, hydromica, bravaisite, and brommallite are names applied to members of this group. The structure is very similar to that of muscovite except that some of the aluminum in the central sheet is partially replaced by magnesium and iron.

6. Hydrated Aluminous Minerals

The minerals gibbsite and diaspore are not considered clay minerals by many workers in the field, but they are included here because they are analogous in structure to the clay minerals and are used in ceramics as clays.

Gibbsite. This mineral has the formula $Al(OH)_3$ and consists of a simple sheet structure as shown in Fig. 2.7. The aluminum is in octahedral coordination with the hydroxyl groups, but only $\frac{2}{3}$ of the available positions are filled. This is the same structure found in the octahedral sheet of the clay minerals. The bonds are weak; thus the mineral is soft and breaks up on heating with relative ease. Gibbsite is a common constituent of soils and clays in warm climates, and is found in a few deposits of high purity.

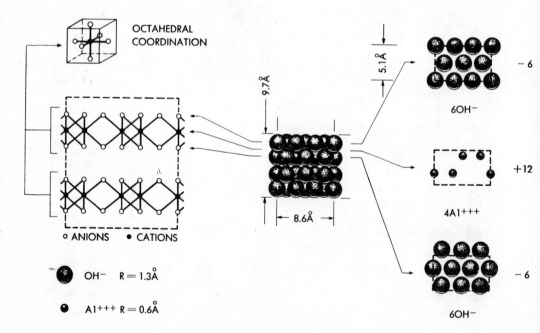

Fig. 2.7   The unit cell of gibbsite, $8Al(OH)_3$.

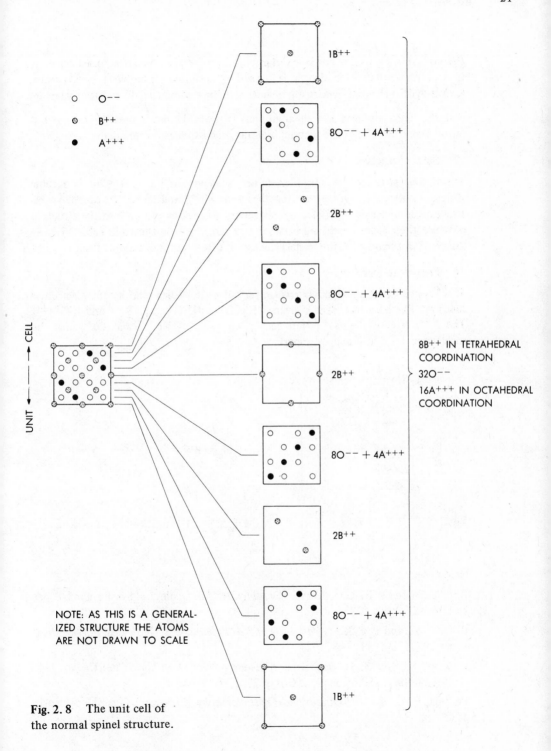

Fig. 2.8 The unit cell of
the normal spinel structure.

Diaspore ($HAlO_2$). This has been an important mineral in the refractories industry. It is probable that the hydrogen is in twofold coordination between two oxygens with a hydroxyl bond making up a sheet structure somewhat like that of gibbsite.

Bauxite. This common ore of aluminum is probably not a specific mineral but rather a mixture of gibbsite, kaolin, limonite, and other minor minerals.

7. Spinel Structure

The crystal structure of a typical spinel is shown in Fig. 2.8. This is a cubic structure with 32 oxygen atoms together with 8 B^{++} and 16 A^{+++} cations. As this structure is unusually flexible, a great variety of cations can go into the structure provided their atomic radii are between 0.6 and 0.9 Å, as shown in Table 17.1. As discussed in Chapter 17, the spinel structure is often found in ceramic stains.

8. Perovskite Structure

This crystal structure is much used in ferroelectric bodies used by the electronics industry. The basic structure is given in Fig. 2.9 with 6 O^{--}, 8 R^{++}, and 1 Ti^{++++}. The R^{++} is generally Ba^{++}, but Ca^{++}, Mg^{++}, Sr^{++}, Rb^{++}, and Cd^{++} may be substituted for it in whole or in part.

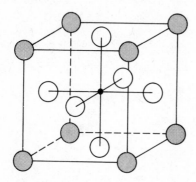

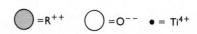

Fig. 2.9   The unit cell of perovskite.

References

Ross, C. S., and P. F. Kerr, *The Kaolin Minerals*, U.S. Geological Survey, Prof. Paper 165, pp.151–180, 1931

Ross, C. S., and P. F. Kerr, *Halloysite and Allophane*, U.S. Geological Survey, Prof. Paper 185-G, pp.135–148, 1934

Ross, C. S., and S. B. Hendricks, *Minerals of the Montmorillonite Group*, U.S. Geological Survey, Prof. Paper 205-B, p.23, 1945

Phillipa, F. C. *An Introduction to Crystallography*, Longmans Green, New York, 1946

Rigby, G. R., The Application of Crystal Chemistry to Ceramic Materials, *Trans. Brit. Ceramic Soc.* **48**, 1, 1949

Kerr, P. F., and P. K. Hamilton, *Glossary of Clay Mineral Names*, Am. Petroleum Institute, Proj. 49, Prelim. Repr. No. 1, 1949

Newham, R. E., A Refinement of Dickite Structure and Some Remarks on Polymorphism in Kaolin Minerals, *Mineralogy Mag.* **32**, 683, 1961

Keeling, P. S., Infrared Absorbtion Characteristics of Clay Minerals, *Trans. Brit. Ceramic Soc.* **62**, 549, 1963

Hinckley, D. N., Variability in "Crystallinity" Values among the Kaolin Deposits of the Coastal Plains of Georgia and South Carolina, *Proc. 11th Nat. Conf. Clays and Clay Minerals*, 299, 1963

Sevratosa, J. M., *et al.*, Infrared Study of the OH Groups in Kaolin Minerals, International Clay Conference, Pergamon, New York, 1963

Keeling, P. S., Investigation of Hydroxyl Groups in Kaolinitic Clays by Infrared Spectrophometer, *Trans. Brit. Ceramic Soc.* **64**, 137, 1965

Grim, R. E., The Clay Mineral Concept, *Bull. Am. Ceramic Soc.* **44**, 687, 1965

Weiss, A., and J. Russow, The Rolling up of Kaolinite Crystals to Halloysite Tubes and Difference Between Halloysite and Tubular Kaolinite, International Clay Conference, Pergamon, New York, 1965

Alietti, A., Identification of Disordered Kaolinites, *Clay Minerals Bull.* **6**, 229, 1966

3

Clay

1. Introduction

Clay has always been the chief ingredient in ceramic bodies from early times, but now many refractory and electronic bodies are made with little or no clay. Nevertheless, clay is still our most important ceramic material. As there are many types of clay, ranging from pure kaolins to shales, it is necessary for the manufacturer to be familiar with those available and their cost at his plant in order to be sure to obtain the best material for his particular purpose. In this chapter the more important clays will be described.

2. Kaolin

Introduction. The word, kaolin, believed to come from the Chinese word *Kaoling,* is applied in this country to most white burning clays. In England, however, the term "china clay" is used. Great quantities of kaolin are mined each year, but only a fifth of it is used in ceramics.

Origin of kaolins. Kaolins may be classified into two types. The first is a residual deposit derived from altered rock, often of the pegmatic type, with the kaolin remaining *in situ* among the undecomposed rock fragments. The second is a sedimentary type in which fine rock and clay particles have been washed out of the original deposit and laid down in lakes and lagoons with some alteration during transport and after settling.

The residual kaolin deposits are usually formed by weathering of pegmatities or mica schists. The reaction may be as follows:

$$KAl\,Si_3O_8 \longrightarrow HAl\,Si_3O_8 + KOH \qquad \text{(hydrolysis)}$$
$$HAl\,Si_3O_8 \longrightarrow HAl\,SiO_4 + 2\,SiO_2 \qquad \text{(desilication)}$$
$$2\,HAl\,SiO_4 + H_2O \longrightarrow (OH)_4\,Al_2\,Si_2\,O_5\ \text{(kaolinite)} \quad \text{(hydration)}$$

It is believed these steps occur by both chemical reaction and by colloidal transfer. The main weathering factor is percolating groundwater containing CO_2, SO_3, and plant acids in solution. In some cases kaolinization takes place because of the action of active fluids rising through the rock mass (pneumatolysis). Large proportions of kaolin are seldom found in residual deposits, since rock fragments often compose 80 to 90% of the whole. However, when the kaolin is removed by washing it is of high purity.

The material in the sedimentary deposit is usually stratified, as the depositing operation causes classification. On the whole, however, these deposits are remarkably uniform over large areas. Along with the kaolin particles are found small amounts of accessory minerals such as muscovite, biotite, quartz, iron oxide, rutile, and garnet. Much of these impurities may be removed by washing.

Deposits of kaolin. The greatest kaolin deposit in the United States is the huge sedimentary band of the Cretaceous age, running from North Carolina down into Alabama just below the Fall Line. The great bulk of our kaolin comes from here. Another deposit of sedimentary nature is found in Putnam County in northern Florida, which has the unique properties of high plasticity and greeen strength. In the Spruce Pine region of North Carolina is found a residual kaolin deposit that yields highly pure material after washing.

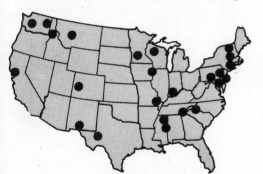

Fig. 3. 1   Residual kaolin deposits in the United States.

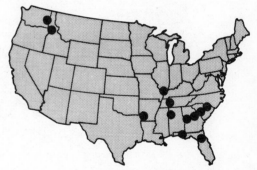

Fig. 3. 2   Sedimentary kaolin deposits in the United States.

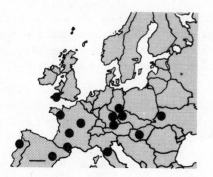

Fig. 3. 3   Residual kaolin deposits in Europe.

The maps in Figs. 3.1 and 3.2 show the reported kaolin deposits in the United States, but only a small proportion of them are being actively worked.

Excellent deposits of kaolin are mined in Cornwall, England, and in France, Germany and Czechoslovakia, as shown on the map of Fig. 3.3.

3. Ball Clays

Introduction. These clays, varying greatly from deposit to deposit, are used in whitewares because of their high plasticity and high green strength. They are never used alone, but always with other clays and nonplastics to make excellent casting slips and highly workable plastic masses. However, they do somewhat impair the whiteness and translucency of porcelain bodies.

Deposits of ball clays. These sedimentary clays are found in a limited area in this country (Fig. 3.4). Most of our supply comes from lower Eocene beds of secondary origin in western Kentucky and western Tennessee. Some of the most famous ball clays come from deposits in the Devon and Dorset areas of southern England.

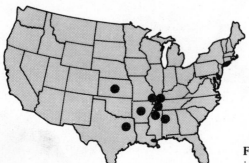

Fig. 3. 4   Ball clay deposits in the United States.

4. Fire Clays

Introduction. These clays are primarily used to make firebrick, refractory mortars, and plastics. They include all clays that are not white-burning and have a fusion point above 1410°C (Cone 15). They may be divided into three classes: 1) flint fire clays having a hard structure, 2) plastic fire clays that have good workability, and 3) high-alumina clays for refractories to withstand unusually severe conditions.

Flint fire clays. We are fortunate in this country to have ample supplies of flint fire clays. These clays are hard, breaking with a concoidal fracture, and develop little plasticity even after grinding. However, when mixed with plastic clays they act as a grog (highly fired clay), thus maintaining the volume stability of the brick. The map in Fig. 3.5 shows that this clay is found in a rather limited area, many deposits being associated with coal seams.

Fig. 3. 5   Flint fire clay deposits in the United States.

Plastic fire clays. These clays, often of Tertiary age, are widespread, as shown on the map of Fig. 3.6. There is considerable variation in the properties of these clays, so they must be carefully selected for a particular purpose. They are used as the. bond in firebrick and as an ingredient in refractory plastics and mortars.

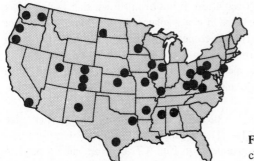

Fig. 3. 6   High heat duty plastic fire clay deposits in the United States.

High-alumina clays. These clays are used for making super-duty and high-alumina refractories. One of these materials, much worked in the past but now nearly exhausted, is diaspore ($HAlO_2$) found only in Missouri. This is a valuable material because of its low shrinkage in firing.

Gibbsite, $Al(OH)_3$, is imported from areas like Dutch Guiana for high-grade refractories. Also our southern bauxite, a mixture of gibbsite and kaolin, is used.

Other clays. Stoneware clays are buff burning with enough fluxes to enable them to fire to a dense structure at temperatures of 1250°C (Cone 8). These clays are used for art pottery and chemical stoneware.

Shales and brick clays are used in great quantities for heavy clay product manufacture. Some are quite plastic, while others must be ground to be workable. All contain enough fluxing materials to allow firing at 1000° to 1200°C (Cone 06–6).

5. Properties of Raw Clays

Introduction. As clay is the chief ingredient in most ceramic bodies, it is important to know as much about its properties as possible in order to understand its functions and possibilities. In this section the more important properties of clays will be discussed.

Particle size. The particle size of clay is a very important characteristic, since it influences many other properties such as plasticity, dry strength, and base exchange capacity.

There are numerous ways of measuring particle size. The microscope and the electron microscope yield absolute size values, but to measure a large number of particles with them is indeed a tedious process. The x-ray method is excellent for particles less than 0.1 micron in size, but requires careful work to measure the resultant broadening of the lines on the diffraction pattern. The most feasible method for the average analysis is the sedimentation method, in which the rate of settling of the particles in water is measured and then converted by Stokes' law into particle size. The assumption must be made that the particles are individual and that their settling rate is the same as that of equivalent spheres. This requires a dilute suspension and complete deflocculation. It has been shown that the platelike clay particles do settle at almost the same rate as a sphere having the same diameter as the width of the plate. Static settling becomes so slow when the particles reach one micron or less in size that thermal currents and Brownian movement tend to introduce serious errors. Therefore measurements below this size are made in a centrifugal field of force. More details about methods can be found in the references at the end of the chapter.

There are numerous methods for plotting the results of the particle size analysis, but it has been found that for clays a simple plot of percent finer versus the logarithm of size is the most convenient. In Fig. 3.7 there are shown particle size curves for a number of ceramic clays. It will be seen that there is a great difference between them. Over half the weight of the particles of Kentucky ball clay have a size below one micron, whereas in Pennsylvania fire clay only one-fifth are below this limit. Many clays have little or no portion below 0.1 micron, while some have several percent in this range. It should be remembered that the most active portion of the clay is in the fine range because of the enormous surface area. Particle size measurements that do not carry down to the finest end of the scale have little meaning in evaluating a clay.

Particle shape. Very little has been known about the exact shape of clay particles, but the electron microscope will be able to give us this information. For example, the kaolinite plates in Fig. 2.4 show the outline clearly, and the shadow at the edge indicates the thickness, which is one-third of the shadow width. In general, kaolins seem to have particles with a thickness of 8 to 10 percent of the width. How closely this holds for wide ranges of particle size or for different kaolin sources is

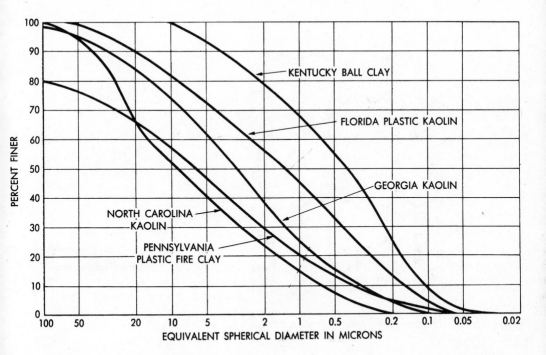

Fig. 3. 7   Particle size distribution for several raw clays.

something that must be determined. A slight change in particle thickness will make a great change in the total surface area of a gram of clay.

Some of the clay minerals, for example halloysite, seem to be made of tubes rolled up from a single, very thin crystal.

Other clay minerals are not as smooth as kaolinite, but are somewhat ragged. Gibbsite particles have actually a rather spongy appearance.

Base exchange capacity. It was shown in Chapter 2 that when a clay mineral such as montmorillonite, with a balanced lattice, had some of its ions replaced by others of different valence (for example, Al^{+++} replaced by Mg^{++}), there would be set up a deficiency of charge in the structure as a whole. This deficiency is balanced by ions adsorbed on the surface of the crystal. However, this is not the only way that ions may be adsorbed, for a balanced lattice like kaolinite can adsorb a small number of ions. It has been thought that this adsorption is due to the broken bonds at the edges of the crystal. This is illustrated diagrammatically in Fig. 3.8, where there are 20 broken bonds for one unit cell of kaolinite.

If a monodisperse (having particles of one size) fraction of kaolinite is available, it is possible to calculate the number of broken bonds per crystal in relation to the ions adsorbed. It will be instructive to make such a calculation for a

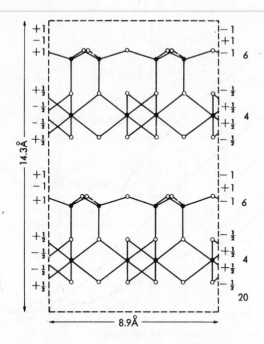

Fig. 3. 8   Schematic diagram of unit cell of kaolinite showing broken bonds.

specific kaolinite particle. The continual rejection of superfluous figures in calculations of this kind should be noted.

The kaolinite crystal is a hexagonal plate 6.0×10^3 Å across the points and the 5.0×10^2 Å thick. It is easy to see that the edge area will be

$$6 \times 3.0 \times 5.0 \times 10^5 = 9.0 \times 10^6 \text{ sq Å.}$$

The volume of each particle will be

$$24 \times 10^6 \times 5.0 \times 10^2 = 12 \times 10^9 \text{ cu Å} = 12 \times 10^{-15} \text{ cc.}$$

The weight of each particle is then

$$12 \times 10^{-15} \times 2.6 = 31 \times 10^{-15} \text{ gm.}$$

Turning to the unit cell shown in Fig. 3.8, we see that the average area of a side is

$$\frac{8.9 + 5.1}{2} \times 14.3 = 100 \text{ sq Å.}$$

The number of broken bonds per unit cell may be seen in the same figure to be 12 from the Si^{++++} and 16 half-strength bonds from the Al^{+++}, making a total of 20, or for the average face, 5.0.

Then the number of broken bonds on the edge of one kaolinite particle will be

$$\frac{9.0 \times 10^6}{100} \times 5.0 = 4.5 \times 10^5.$$

The base exchange capacity of one gram of this kaolinite is found by experiment to be 1.6×10^{-5} equivalents per gram. Multiplying by Avogadro's constant, 6.0×10^{22}, and by the weight of one particle gives the number of ions adsorbed on one particle:

$$1.6 \times 10^{-5} \times 6.0 \times 10^{23} \times 31 \times 10^{-15} = 3.0 \times 10^{5}.$$

This close agreement is interesting but not conclusive, for some of the ions may be adsorbed on the face of the crystal, and some of the edge positions may be vacant.

The maximum capacity to adsorb ions is called the base exchange capacity and is expressed in milliequivalents per 100 grams of clay.

There are a number of methods of measuring this quantity, but in ceramics it is convenient to electrodialyze the clay slip in a three-compartment cell, as shown in Fig. 3.9. In this way the adsorbed ions are stripped off and replaced by hydrogen. If the hydrogen-clay slip is then titrated with a base such as NaOH, and the pH (hydrogen ion concentration) of the slip measured, a curve like that in Fig. 3.10 will be obtained. The inflection point in this curve will represent the base exchange capacity.

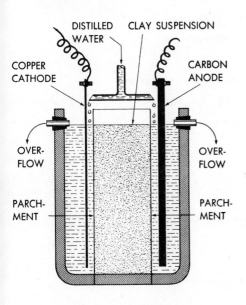

Fig. 3. 9 Three-compartment electrodialysis cell.

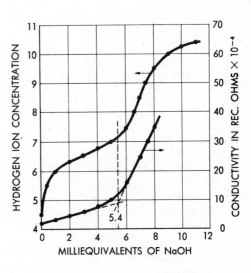

Fig. 3.10 Two methods of measuring the base exchange capacity of a clay.

In Table 3.1 are listed the base exchange capacities for some monodisperse fractions of kaolinite and for a few complete clays. For kaolinite, the base exchange capacity is closely proportional to the total surface area.

Table 3.1   Base exchange capacity

Clay	Mean spherical diameter in microns	Surface area per 100 gm/m^2	Base exchange capacity in milli-equiv. per 100 gm
Kaolinite fraction	10	1.1	0.4
" "	4	2.5	0.6
" "	2	4.5	1.0
" "	1	11.7	2.3
" "	0.5	21.4	4.4
" "	0.2	39.8	8.1
Georgia kaolin			1
Pa. flint fire clay			5
Ky. flint fire clay			7
Plastic fire clay			7
Ky. ball clay			12
Bentonite			100

Table 3.2   Separation of accessory minerals in clay

Mineral	Specific gravity	Heavy liquid
Gypsum	2.3	
Gibbsite	2.4	2.5-Tetrabromoethane + carbon tetrachloride
Orthoclase	2.6	
Microcline	2.6	
Albite	2.6	(Clay minerals in this group) 2.65-Tetrabromoethane + carbon tetrachloride
Quartz	2.7	
Calcite	2.7	
Anorthite	2.8	
Biotite	2.8	
Muscovite	2.8	
Beryl	2.8	
Tourmaline	3.0	2.9-Tetrabromoethane + carbon tetrachloride
Magnesite	3.0	
Garnet	3.4	
Limonite	3.8	
Corundum	4.0	3.87-Sol. of thallium formiate at 40°C
Ilmanite	4.2	
Rutile	4.5	
Zircon	4.7	
Pyrite	5.0	
Hematite	5.2	
Magnetite	5.2	

As will be shown in Chapter 8, the base exchange capacity is an important property of casting slips.

Accessory minerals. Natural clays contain many types of crystalline matter, and those that are not clay minerals are known as accessory minerals. Often these accessory minerals are of great importance in evaluating the worth of a clay. They can be determined most readily by centrifuging the clay in a series of heavy liquids to separate the various minerals in groups according to density. A suitable division is shown in Table 3.2. After each group is separated, it may be examined under the petrographic microscope or by x-ray diffraction methods. There the species and relative amounts of the minerals can readily be determined, since generally there are only two or three minerals in each group.

The more common accessory minerals in clays are quartz, feldspars, micas, and the iron minerals.

Organic matter. All clays contain some organic matter. There is very little in the residual kaolins but a large amount in ball clays. The organic matter is in the form of lignite, waxes, or humic acid derivatives, as described by Worral (1956). The organic matter undoubtedly has an important influence on the plastic and dried properties of clays.

In studying the properties of the clay minerals it is necessary to remove the organic matter without destroying the crystal structure. This may most readily be done by digestion in a hydrogen peroxide solution. A long time is required to remove all organic matter; several months, with frequent renewal of the peroxide, is often necessary.

Color. The color of the raw clay is of little importance in ceramics, since the heat used will destroy or alter it. For the paper trade, however, color is of great importance, and it is possible to obtain an exact measure of it by means of the recording spectrophotometer.

Chemical composition. A knowledge of the chemical composition of clays is helpful in evaluating them for a specific use. However, this information must be used in combination with the physical properties to obtain a complete picture.

Table 3.3 lists the chemical analyses of a considerable number of typical clays and may be used as a reference when studying new clays.

Unfortunately, only a few analyses are available which are really complete, so we have little idea of the minor constituents in many types of clay. In general, it will be seen that the less pure the clay, the lower will be the amount of combined water. The greater the amount of flux ($CaO + MgO + K_2O + Na_2O + Fe_2O_3$), the lower will be the maturing temperature. The North Carolina residual kaolin is one of the few American clays with the low TiO_2 content of the English china clays, but the latter contain slightly more feldspar.

Table 3.3 Chemical analyses of typical clays

Constituent	English china clay, washed	N.C. kaolin, washed	Zettlitz kaolin, washed	Georgia sedimentary kaolin	Ball clay, Mayfield, Ky.	Ball clay, Tenn.[1]	Gibbsite, Dutch Guiana	First-grade diaspore, Mo.	Siliceous clay, Rush, Tex.[2]
SiO_2	48.3	46.18	46.87	45.8	56.4	53.96	4.5	10.9	82.45
Al_2O_3	37.6	38.38	38.00	38.5	36.00	29.34	58.4	72.4	10.92
Fe_2O_3	0.5	0.57	0.89	0.3	(see above)	0.98	3.2	1.1	1.08
FeO									
Fe_2S									
MgO	0.1	0.42	0.35		tr.	0.30			0.96
CaO	0.1	0.37	tr.	tr.	0.4	0.37	0.4		0.22
Na_2O	0.1	0.10	1.22	0.3	2.0	0.12			
K_2O	1.9	0.58		0.08	3.3	0.28			
H_2O									
H_2O+	12.0	13.28	12.70	13.6	7.9	12.82	30.6	13.5	2.40
CO_2									
TiO_2	0.05	0.04		1.4		1.64	2.9	3.2	1.00
P_2O_5						.15			
SO_3						.03			
MnO						.02			
ZrO_2		0.08							
Org. C									
Org. H									

Table 3.3 (*continued*)

Constituent	Brick shale, Mason City, Ia.[2]	Brick clay, Milwaukee, Wis.[2]	Glacial brick clay, Boston, Mass.[2]	Flint fire clay, Cambria, Pa.[3]	Flint fire clay, Carter, Ky.[3]	Semi-flint fire clay, Clearfield, Pa.[3]	Plastic fire clay, Lawrence, O.[3]	Diaspore fire clay, Mo.	Burley flint fire clay, Mo.
SiO_2	54.64	38.07	57.02	44.43	44.78	43.04	58.10	29.2	33.8
Al_2O_3	14.62	9.46	19.15	37.10	35.11	36.49	23.11	53.3	49.4
Fe_2O_3	5.69	2.70	6.70	0.46	1.18	1.37	1.73	1.9	1.9
FeO				0.55	0.74	0.83	0.68		
Fe_2S				0.22	0.14	0.24	0.55		
MgO	2.90	8.50	3.08	0.19	0.55	0.54	1.01		
CaO	5.16	15.84	4.26	0.60	0.77	0.74	0.79		
Na_2O	5.89 }	2.76 }	2.83	0.10	0.29	0.46	0.34		
K_2O			2.03	0.55	0.44	1.10	1.90		
H_2O	0.85			0.80	0.84	0.82	2.27		
H_2O+	3.74	2.49	3.45	12.95	13.07	12.44	7.95	12.0	12.0
CO_2	4.80	20.46		0.71	0.07	0.05	0.05		
TiO_2			0.91	1.84	2.22	1.79	1.40	2.7	2.6
P_2O_5				0.21	0.02	0.10	0.17		
SO_3				0.01	0.01	0.01	0.03		
MnO				0.01	0.02	0.01	0.01		
ZrO_2				0.01	0.01	0.01	0.01		
Org. C				0.10	0.11	0.22	0.22		
Org. H						0.03	0.03		

[1] Spinks Clay Company.
[2] Reis, *Industrial Minerals and Rocks.*
[3] Downs Sheraf, analyst.

6. Plastic Properties

There has never been developed a particularly good quantitative test for plasticity. However, there are certain properties contributing to plasticity that are subject to precise measurement, such as the yield point and extensibility, which will be discussed in Chapter 7. We do not have any reliable values for typical clays, and consequently our discussion must be rather general.

The finer-grained clays are highly plastic, but even coarse-grained clays, containing a small portion of montmorillonite, may be quite plastic. On the other hand, shales and flint clays require fine grinding to develop this property. Clays containing appreciable amounts of accessory minerals such as sand lose plasticity.

7. Dried Properties

Drying shrinkage. This property is readily measured by determining either the length or volume change when clay is dried, as discussed in Chapter 9. This property is of great importance when forming large pieces, for a high shrinkage necessitates very slow drying to prevent cracking. In general, the fine-grained, plastic clays have the higher shrinkage values.

Dried strength. This property is important to facilitate handling ware between the dryer and the kiln. Again the fine-grained clays, especially those containing montmorillonite, are the strongest.

There has been much speculation about the cause of dry strength, but the most reasonable explanation seems to be the van der Waals forces between the flat faces of the clay particles as they come together, somewhat as shown in Fig. 7.11. It will be noted that the forces are sufficient to orient the plates so that the edges are substantially parallel.

8. Slaking Properties

The time required for a one-inch cube of dry clay to disintegrate after being immersed in water is usually taken as a measure of slaking. This property varies a great deal; flint clays and some shales take an infinitely long time, whereas washed North Carolina kaolin slakes in 10 minutes. The slaking time has considerable bearing on the process and equipment needed to break down a raw clay to the plastic state.

9. Firing Properties

Since the firing properties, such as shrinkage, porosity, and many other factors, are discussed in Chapter 10, no reference will be made here to these important characteristics of clay.

References

Sortwell, H. H., *American and English Ball Clays*, National Bureau of Standards T.P. no. 227, 1923

Norton, F. H., and S. Speil, The Measurement of Particle Sizes in Clays, *J. Am. Ceramic Soc.* **21**, 89, 1938

Worrall, W. E., The Ceramic Organic Matter in Clays, *Trans. Brit. Ceramic Soc.* **55**, 689, 1956

Clark, N. O., Control of China Clays, *J. Brit. Ceramic Soc.* **1**, 262, 1964

Clarke, O. M., Clay Deposits of the Tuscaloosa Group in Alabama, *Proc. 12th Nat. Conf. Clays and Clay Minerals* **19**, 495, 1964

Bates, T. F., Geology and Mineralogy of the Sedimentary Kaolins of the Southwestern United States, *Proc. 12th Nat. Conf. Clays and Clay Minerals* **19**, 177, 1964

Keeling, P. S., The Nature of Clay, *J. Brit. Ceramic Soc.* **2**, 236, 1965

Grim, R. F., *Clay Mineralogy*, 2nd ed., McGraw-Hill, New York, 1968

Norton, F. H. *Fine Ceramics*, McGraw-Hill, New York, 1970

4

Feldspars and other fluxes

1. Introduction

Whiteware bodies require a flux to supply at least a portion of the glassy phase during firing which is needed for strength and translucency. The flux most commonly used is feldspar, as it is plentiful, low in iron, and insoluble in water. In fact it is one of our few sources of insoluble alkalis. This latter property also makes it useful in raw glazes and glass batches. Feldspars are quite variable in composition, but generally conform to the expression

$$K_x Na_{1-x} \begin{bmatrix} Al \\ Si_3 \end{bmatrix} O_8.$$

2. Feldspar Minerals

Commercial feldspars are found in pegmatites along with quartz and mica. The pegmatites are usually found in dikes (fissures in older rocks later filled with magma). Because seams of mica are often found along the dikes, many of the older feldspar mines in New England were previously mica mines.

The three common feldspar minerals are:

$$\text{Orthoclase} \quad K \begin{bmatrix} Al \\ Si_3 \end{bmatrix} O_8$$

$$\text{Albite} \quad Na \begin{bmatrix} Al \\ Si_3 \end{bmatrix} O_8 \left.\vphantom{\begin{bmatrix} Al \\ Si_3 \end{bmatrix}}\right\} $$

$$\text{Anorthite} \quad Ca \begin{bmatrix} Al_2 \\ Si_2 \end{bmatrix} O_8 \quad \text{Plagioclase}$$

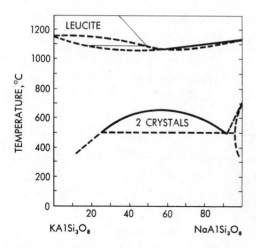

Fig. 4. 1   Equilibrium diagram for potash and soda feldspars.

The plagioclase feldspars form a continuous series, but the orthoclase with large K^+ ions forms only a limited series with albite (Fig. 4.1). The commercial feldspars belong to this latter series.

3. Occurrence of Feldspars

The map of Fig. 4.2 shows the location of the important deposits of feldspar in the United States. At present North Carolina is much the largest producer.

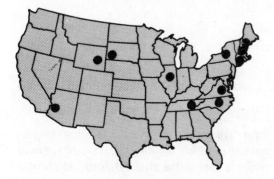

Fig. 4. 2   Feldspar deposits in the United States.

4. Mining and Milling Feldspar

In earlier times the crude rock was mined by local farmers and brought in to a central mill. Here it was hand sorted and ground to produce a uniform product. Now large mills receive the crude rock from quarrying operations and then grind and treat it by froth floatation methods to remove quartz and mica.

Table 4.1 Chemical and mineral analyses of some feldspars

Constituent	North Carolina	North Carolina	Maine	Canada	Georgia
SiO_2	69.5	70.0	67.8	65.5	65.0
Al_2O_3	17.5	18.1	18.4	18.7	19.5
Fe_2O_3	0.1	0.1	0.1	0.1	0.05
CaO	0.8	1.5	0.3	0.4	0.2
MgO	tr.	tr.	tr.	tr.	tr.
K_2O	8.1	3.5	10.0	12.8	13.1
Na_2O	3.6	6.5	3.0	2.3	2.1
Loss	0.3	0.3	0.3	0.2	0.3
Potash feldspar	47.9	20.7	59.2	75.7	77.5
Soda feldspar	30.6	55.2	25.4	19.5	17.5
Lime feldspar	4.0	7.5	1.5	2.0	1.0
Quartz	14.1	15.7	8.8	1.2	2.5
Other minerals	3.4	0.9	5.1	1.6	1.5

5. Composition of Feldspars

In Table 4.1 are shown the chemical analyses of a number of commercial feldspars. It will be noted the ratio of soda to potash varies considerably. Generally the high-potash feldspars are preferred for bodies and those high in soda are used in glass, glazes, and enamels.

6. Other Fluxing Materials

Although feldspar is the principal ceramic flux, there are others which are quite as important for some uses.

Nepheline seyenite. This mineral contains about 50% albite, 25% microcline $(K \begin{bmatrix} Al \\ Si_3 \end{bmatrix} O_8)$, and 25% nepheline $(Na_2 \begin{bmatrix} Al_2 \\ Si_2 \end{bmatrix} O_8)$. Thus it has a somewhat higher alumina and soda content than feldspar. The most important deposit is in Ontario, where the rock is crushed and ground after which the iron minerals are removed with a magnetic separator. This material is used in the glass batch to add alumina and in whiteware bodies to lower the maturing temperature. A typical analysis is shown in Table 4.2.

Graphic granite. This is a rock containing about 75% feldspar and 25% quartz. It is found in large, uniform deposits and should be a future source of feldspar.

Cornish stone. This is a flux much used in England. It is a partially decomposed pegmatite containing albite, orthoclase, and small amounts of kaolin and fluorides. It is mined and treated by the processes used for feldspar. Recently a grade has been made available with most of the fluorides removed (see Table 4.2).

Table 4.2 Chemical analyses of other fluxing materials

Constituents	Nepheline syenite	Graphic granite	Cornwall stone	Defluorinated Cornwall stone
SiO_2	60.2	72.4	76.2	79.5
Al_2O_3	23.7	14.5	12.1	12.4
Fe_2O_3	0.1	0.3	0.2	0.1
CaO	0.4	0.2	1.4	−
MgO	0.1	−	0.1	−
TiO_2	−	−	0.5	0.1
K_2O	5.0	10.1	4.6	3.8
Na_2O	10.0	2.2	3.7	3.9
F	−	−	1.1	0.1
Loss	0.5	0.3	0.1	0.1
Total	100.0	100.0	100.0	100.0

Bone ash. This flux, the important ingredient of bone china, is produced by calcining and grinding ox bones. When the calcination is correct a little colloidal organic matter is left, which enhances the working properties; however, the main constituent is calcium phosphate, $3\,CaO \cdot P_2O_5$.

Flurorite (CaF_2). This flux, the main constituent of flurospar, is used in enamels, glasses, and glazes. Cryolite ($3NaF \cdot AlF_3$) is also used to introduce fluorine. Lead fluoride and sodium or potassium silico-fluoride are used for the same purpose. Spodumene ($Li_2O \cdot Al_2O_3 \cdot 4SiO_3$) is a convenient material for introducing lithium into the batch.

References

Omeara, R. G., and J. E. Norman, Froth Floatation and Agglomerate Tabling of Feldspars, *Bull. Am. Ceramic Soc.* **18**, 286, 1939

Dubois, H. B., Development and Growth of the Feldspar Industry, *Bull. Am. Ceramic Soc.* **19**, 206, 1940

Bowles, D., and C. W. Justice, Feldspar in a Period of Change. U.S. Bureau of Mines, Inf. Circular 2287, 1947

Schairer, J. F., The Alkali-Feldspar Join in the System $NaAlSiO_4 - KAlSiO_4 - SiO_2$, *J. Geol.* **58**, 512, 1950

Keeling, P. S., Cornish Stone, *Trans. Brit. Ceramic Soc.* **60**, 390, 1961

Reeves, J. E., Nepheline Syenite, Canada Dept. of Mines Bull., 1962

Minerals Yearbook, U.S. Bureau of Mines, Government Printing Office, Washington, D.C., 1964

5

Other minerals

1. Introduction

In addition to the clays and fluxing materials, there are many other minerals used in the ceramic industry. These will be discussed in this chapter.

2. Silica (SiO_2)

Quartz, the principal mineral of silica, is both wide-spread and abundant with a high degree of purity. Other forms such as cryptocrystalline silica (flint) and diatomite are useful in ceramics. Quartz is used in whiteware bodies, glass, glazes, enamels, silica brick, and sand-lime brick.

Quartz. This mineral has been studied very thoroughly and therefore its properties are well known (see Table 5.1). Quartz is one of the few minerals found in large, optically perfect crystals. From these may be cut specimens for optical instruments and for high-frequency electrical oscillators.

The atomic structure of quartz has been shown to consist of a three-dimensional network of SiO_4 tetrahedrons linked into a compact structure, as would be expected from its high specific gravity. The open holes in the structure are so small that other atoms cannot enter and therefore the crystals are always of high purity.

One of the most characteristic properties of quartz is the reversible inversion from the low to the high form at a temperature of $573°C$, which will be discussed in Chapter 10.

Quartz occurs as an important constituent of some igneous rocks, such as granite and diorite, and is a less common constituent of other igneous rocks. It is found in most metamorphic rocks, comprising the major portion of the sandstones and occurring in smaller but definite amounts in clays and shales. Quartz in a pure form is often found in veins running through other rocks.

Table 5.1   Properties of quartz

Property	Value
Specific gravity	2.651
Hardness (Knoop)	820
Melting point ^{o}C	1728
Specific heat (0 - 200°C)	0.203
Coefficient of expansion $\parallel$ (°C)	7.5×10^{-6}
Coefficient of expansion $\perp$ (°C)	13.8×10^{-6}
Index of refraction	1.544
Birefringence	0.009
Crystal system	Hexagonal
Cleavage	Difficult
Dielectric constant	4.5

Crystalline quartz is widespread in nature and many deposits are found in this country. However, good optical crystals are not plentiful, and almost all of these have been imported from Brazil. As quartz crystals can now be grown in the laboratory, we are not as dependent on this source as formerly. Rock quartz is not used to any great extent in ceramics because of the cost of grinding and of removing the resultant iron contamination.

The principal source of quartz for the ceramic industry is sandstone consisting of lightly bonded quartz grains. For some uses this sandstone is simply disintegrated, but for others it is readily ground in pebble mills with no contamination. In this country we are fortunate in having excellent deposits of these sandstones of high purity. For the glass batch, these sandstones must not only be of high purity, but must also disintegrate readily to a rather uniform grain size. Two types of sandstone fit these conditions, the Oriskany sandstone found from New England to Alabama, and the St. Peter sandstone in Illinois and Missouri. Important deposits are shown in Fig. 5.1. Table 5.2 shows the analysis of some of

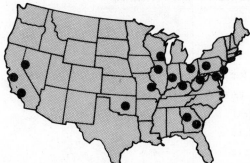

Fig. 5. 1   Deposits of glass sand in the United States.

Table 5.2   Analyses of glass sands

Constituent	Mapleton, Pa.	Hancock, W. Va.	Ottawa, Ill.
SiO_2	99.82	99.81	99.61
Al_2O_3	0.12	0.17	0.16
Fe_2O_3	0.017	0.014	0.021
CaO	tr.	0.00	0.050
MgO	tr.	0.00	0.03
Ig. loss	—	—	0.08

Table 5.3   Specifications for chemical composition of glass sands[1]

Qualities	SiO_2 min.	Al_2O_3 max	Fe_2O_3 max	CaO + MgO max
1st quality (optical glass)	99.8	0.1	0.02	0.1
2nd quality (flint containers and tableware)	98.5	0.5	0.035	0.2
3rd quality (flint glass)	95.0	4.0	0.035	0.5
4th quality (sheet and plate glass)	98.5	0.5	0.06	0.5
5th quality (sheet and plate glass)	95.0	4.0	0.06	0.5
6th quality (green glass containers and window glass)	98.0	0.5	0.3	0.5
7th quality (green glass)	95.0	4.0	0.3	0.5
8th quality (amber glass containers)	98.0	0.5	1.0	0.5
9th quality (amber glass)	95.0	4.0	1.0	0.5

[1]Recommended by the American Ceramic Society and the National Bureau of Standards.

Table 5.4 Screen analyses of some glass sands

Mesh size	Mapleton, Pa. (per cent retained)	Ottawa, Ill. (per cent retained)	Berkeley Springs, Va. (per cent retained)
14	0	—	—
20	0.5	0	—
28	3.8	3.4	0
35	17.5	30.5	1.1
48	56.9	65.0	33.6
65	90.9	82.1	78.1
100	98.5	92.9	94.6
150	—	98.2	98.6

these deposits, Table 5.3 the chemical specifications for glass sand, and Table 5.4 their natural grain sizes.

Quartzites for silica brick, called ganister, are very firmly consolidated sandstone, so that fracture occurs across the grains and thus permits crushing into fragments of the desired size. These quartzites are found in the Medina formation in Pennsylvania, the Baraboo formation in Wisconsin, and formations in Alabama and Colorado, as well as a few other places (Fig. 5.2).

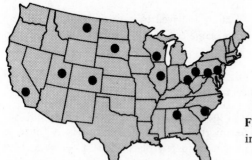

Fig. 5. 2   Deposits of ganister in the United States.

Naturally occurring sands are widespread, but are seldom sufficiently pure for ceramic use. They are employed as foundry sand, as cement aggregate, for sand blasting, for grinding glass, and for many other uses.

With reference only to the ceramic industry, the uses of quartz may be roughly divided as shown in Table 5.5.

Table 5.5 The use of quartz in the ceramic industry

Type of use	Percent of total
Glass sand	83
Ganister for silica brick	12
Whiteware bodies	4
Whiteware glazes	0.5
Enamels	0.5

Cristobalite. This is another form of silica. It is less pure than quartz, and some geologists feel that these impurities have prevented the formation of the quartz crystal. Cristobalite is not plentiful in nature and is of little use as a source of silica, but as a constituent of fired ceramics it is important and will be discussed in Chapter 10.

Tridymite. This form of silica is rare in nature but again is important in fired ware.

Vitreous silica. This glassy material, the result of a lightning flash fusing sandy soil, is found in nature in only a few places. As a manufactured product, however, it is of great value because of its very low coefficient of expansion.

Mircocrystalline forms. In Europe the silica used in whiteware bodies is flint from calcined flint pebbles found in the chalk beds. The term "potter's flint" has been carried over to this country, even though we use pulverized quartz almost entirely. As far as the final product is concerned, it seems to make little difference which form of silica is used.

The microcrystalline kinds of silica are believed to have formed at relatively low temperatures. There are many types, some of fibrous habit and all very finely crystalline. Most of them contain some water.

Hydrated silica. Opal is typical of this class. This form is an amorphous silica gel with some water. It is of little use in ceramics.

Diatomite. This form of silica consists of the skeletons of diatoms about 10 microns in diameter. The silica is believed to be amorphous. This material is widely found in nearly every bog in the country, but in only a few places does it occur in thick enough layers and of sufficient purity to make it worthwhile to mine it. Commercial deposits are shown in Fig. 5.3. The best material comes from the West Coast.

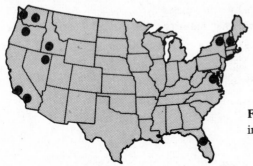

Fig. 5.3   Deposits of diatomite in tne United States.

Diatomite is used largely in heat insulators, as the porous skeletons provide myriads of fine pores.

3. Magnesite

Uses. This basic material, when converted to MgO, is an important refractory because of its high fusion point and excellent resistance to slags. To a lesser extent it is used in bodies and glazes. Fused magnesia is an important high-temperature electrical insulator.

Origin. Magnesite occurs in two main forms: the crystalline and the dense or cryptocrystalline. The crystalline variety is usually formed in nature by the alteration of dolomite through the percolation of magnesium solutions. The dense variety is often the alteration product of serpentine types of rock.

Ocurrence. There is almost no high-quality magnesite mined in the United States, but much pure cryptocrystalline material is being imported from India and Greece. However, the refractories industry has depended largely on magnesia extracted from sea water. The process, shown in the flow sheet of Fig. 6.13, depends on an ion-exchange process. As sea water contains only 0.6% of magnesium compounds, great quantities of sea water must be handled. There are now about a dozen plants producing sea-water magnesite in this country.

Burning. Magnesite for refractories is dead-burned at a temperature of 1600 to 1900°C in shaft or rotary kilns to a dense periclase (MgO) structure. For electrical insulation purposes the magnesite may be fused in electric arc furnaces at temperatures of over 3000°C.

4. Lime (CaCO$_3$)

Ample supplies of limestone are available in this country. The lime rock is calcined in shaft or rotary kilns to produce quicklime (CaO), which is then slaked to make lime hydrate (Ca(OH)$_2$), used in mortars and plasters. When high-purity calcium carbonate is required for use in glazes, enamels, or optical glass, it is produced by chemical means and is referred to as whiting.

5. Dolomite

This rock is a mixture of magnesite and calcite crystals in varying proportions. There are ample deposits in the United States of this material, which is used in basic refractories, especially for the oxygen converter.

6. Chromite

This mineral is a spinel, Cr$_2$FeO$_4$, although much of the Cr^{+++} and the Fe^{++} may be replaced in the lattice to leave a balanced structure as

$$\left.\begin{array}{c} Cr_2O_3 \\ Al_2O_3 \end{array}\right\} \left\{\begin{array}{c} FeO \\ MgO \end{array}\right.$$

and for many purposes these replacements are not harmful. However, silica will not go into this spinel lattice and so it acts as an undesirable impurity. Sometimes MgO is added to form with the silica the stable mineral forsterite (Mg$_2$SiO$_4$), thus tying up the undesirable silica.

Origin. Chromite ores are believed to originate both from magmatic crystallization and by precipitation from hydrothermal solutions. Mining is done by the usual

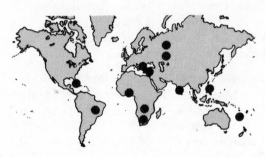

Fig. 5.4   World deposits of cnrome ore.

methods, with hand sorting to remove the gangue. Concentration methods such as flotation have not as yet been used to any extent.

Unfortunately, the United States has very limited sources of chrome ore, and must therefore import nearly all the ore used here. Chrome ore is mined in Turkey, Rhodesia, New Caledonia, the Philippines, Cuba, the U.S.S.R., and a few other places, as shown on the map in Fig. 5.4. Typical analyses are given in Table 5.6.

Table 5.6 Chemical analyses of chrome ores

Constituent	Turkey	Rhodesia	Phillippines	New Caledonia	Russia	Cuba
Cr_2O_3	46.6	45.4	32.1	54.5	46.2	30.5
SiO_2	6.7	7.5	5.3	3.1	4.0	6.1
Al_2O_3	12.5	13.8	27.6	11.0	14.6	27.5
FeO	12.9	15.1	13.0	19.5	15.6	14.2
CaO	1.2	0.5	1.1	1.5	0.3	0.9
MgO	17.3	13.6	18.2	8.0	15.4	18.3
Total	97.2	95.9	97.3	97.6	96.1	97.5

Uses. Chromite is used in ceramics largely as a refractory in the form of burned bricks, chemically bonded bricks, and as plastics. For this purpose a low-silica material is desired. Chromite is also a source of chromic oxide and other compounds used as a color or stain.

7. Silicate Minerals

Talc. The specific mineral is $(OH)_2Mg_3(Si_2O_5)_2$, but variations occur by ion replacement and by variations in the lattice spacings. Steatite is a rock consisting largely of talc.

The talc crystal is a three-layer structure like montmorillonite but with all the octohedral positions filled with Mg^{++} (brucite sheet). Theoretically it contains 63.5% SiO_2, 31.7% MgO, and 4.8% H_2O. The weak bonding between sheets gives easy cleavage and softness.

Fig. 5. 5 Deposits of talc in the United States.

The United States talc deposits are shown on the map of Fig. 5.5 and the composition of the various deposits is given in Table 5.7. Most of the talc is taken from underground mines, crushed, and sized or it may be further treated by flotation to remove quartz. Ceramic talc is ground to pass a 200-mesh screen. For high-frequency use talc must have less than 2% CaO, less than 0.3% Fe_2O_3, and less than 0.3% $(Na_xK_{1-x})_2O$.

Talc is used in wall tile bodies very extensively and in all steatite bodies.

Table 5.7 Chemical analyses of talcs and pyrophyllite

Constituer	New York[1]	Vermont[1]	North Carolina[1]	California[1]	Georgia[1]	Pyrophyllite[2]	Wollastonite
SiO_2	66.2	56.3	61.4	59.6	41.0	63.5	50.9
Al_2O_3	1.1	3.2	4.4	1.7	4.2	28.7	0.7
Fe_2O_3	0.6	5.4	1.7	0.9	5.9	0.8	0.3
MgO	25.7	27.9	26.0	30.0	28.6	tr.	0.2
CaO	2.3	0.4	0.8	0.8	4.8	tr.	47.6
MnO	0.2	0.1	—	—	—	—	0.1
$K_2O + Na_2O$	—	0.9	—	0.3	—	0.4	—
CO_2	0.6	0.4	—	—	—	—	—
Comb. water	3.9	5.7	5.1	5.9	15.5	5.9	—

[1] From *Industrial Minerals and Rocks*.
[2] From *Ec. Pap.* No. 3, N.C. Geol. Surv.

Pyrophyllite. This mineral is much like talc except that the Mg^{++} is replaced by Al^{+++}, giving a soft but refractory structure. Its main use is in tile bodies and refractories. Our most important source is in North Carolina. A typical analysis is shown in Table 5.7.

Wollastonite. This is a relatively pure calcium silicate ($CaO \cdot SiO_2$) and is used in low-loss dielectric bodies and for wall tiles. The most important deposit is in Willsboro, N.Y. Its composition is shown in Table 5.7.

8. Aluminous Minerals

These minerals are used in refractories and are a source of alumina.

Gibbsite ($Al(OH)_3$). There are no important deposits of this mineral in the United States, but considerable amounts are imported from Dutch Guiana. The bauxites found in our southern states are considered mixtures of kaolin and gibbsite, but are not pure enough for many ceramic uses.

Diaspore ($HAlO_2$). This mineral is found in isolated pockets in Missouri, where it has been mined for the refractories industry. At present the supply is practically exhausted.

Sillimanite minerals (Al_2SiO_5). The three minerals, sillimanite, kyanite, and andalusite, all have the formula (Al_2SiO_5), which gives 37% silica and 63% alumina. On heating they break down to mullite and silica. There are a few deposits of these minerals in the United States but most of the material is imported from India for use in refractories and special porcelains.

Corundum (Al_2O_3). There are many deposits of impure corundum (known as emery) in this country, which are mined for loose and coated abrasives. A small amount of massive corundum has been imported from Africa for use in grinding wheels. Of course, large amounts of corundum are made in the electric furnace.

9. Pure Oxides

An increasing number of pure refractory oxides are now used in bodies having specific electrical and thermal properties.

Alumina (Al_2O_3). This oxide, prepared by the Bayer process from bauxites, is now used in considerable quantities for spark-plug cores, valve seats, electronic substrates, and many other purposes. Table 5.8 gives the characteristics of some commercially available material.

Beryllia (BeO). This oxide is used for heat sinks in electronic circuits because of its very high thermal conductivity. It can be obtained in purities up to 99.85% BeO.

Tin oxide (SnO_2). This material is used for electrodes in glass furnaces and would serve as an excellent refractory if it were not so costly. All tin ores must be imported.

10. Carbonaceous Materials

Graphite. This is a black, flaky mineral consisting entirely of carbon. The atomic structure (Fig. 5.6) consists of widely spaced layers of hexagonally packed carbon atoms, which accounts for its softness and flakelike character. It is found in nature in a number of places such as Ceylon, Madagascar, and Korea. It may also be formed from coke in the electric furnace. Graphite is of great value as a refractory because of its high electrical and thermal conductivity and its high stability in a

Table 5.8 Properties of aluminas used in fine ceramics[*]

Characteristic	Type of alumina					
Number	A-2	A-3	A-5	A-10	A-14	T-61
Characteristic	Calcined	Calcined	Calcined	Low soda	Low soda	Tabular alpha
Al_2O_3, %	98.9	99.2	99.2	99.5	99.6	99.5+
SiO_2, %	0.02	0.02	0.02	0.08	0.12	0.06
Fe_2O_3, %	0.03	0.03	0.03	0.03	0.03	0.06
Na_2O, %	0.45	0.45	0.50	0.2	0.04	0.02
Ignition loss, %	0.6	0.4	0.2	0.3	0.3	0
Size analysis, cum, %						
On 100 m	4-15	4-15	2-10	4-15	4-15	
On 200 m	50-75	50-75	40-65	50-75	50-75	
On 325 m	88-98	88-98	75-95	88-98	88-98	
Through 325 m	2-12	2-12	5-25	2-12	2-12	
Bulk density, packed lb/ft³	68	68	63	80	83	135
Firing shrinkage, linear %	19-23	20-22	16-18	14-16	16-18	10

[*]Data in this table were supplied by the Aluminum Company of America. Similar products are made by Reynolds, Alcan, and Kaiser in the United States and firms in England, Germany, Japan, and France.

The calcined aluminas are used in bodies for high strength, good abrasion resistance, and chemical inertness. The low-soda aluminas are used in electrical bodies demanding a low loss factor. The tabular alumina is used in refractories and electrical insulators.

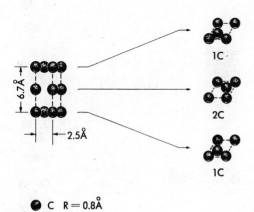

$\bullet$ C R = 0.8Å

Fig. 5.6 Unit cell of graphite showing layer structure.

reducing atmosphere. Mixed with clay, graphite makes excellent crucibles for metal melting. It also has nuclear properties that make it valuable in a reactor.

Coal and coke. Coal is a widely distributed fuel that may be changed to coke by distilling off the volatile matter. The coke may be finely divided and pressed into a solid mass by means of an organic binder which is later carbonized by heating in a reducing atmosphere. Carbon is used for electrodes, crucibles, refractory blocks, and many other purposes involving high-temperature service.

Diamond. This extraordinary mineral is a form of carbon with a hardness far above that of any other known material, as shown in Table 5.9.

Table 5.9 Hardness values for some minerals[*]

Material	Knoop hardness
Diamond	8000
Boron carbide	2800
Silicon carbide	2500
Corundum	2000
Topaz	1300
Quartz	800

[*]Thibault, N. W., and Nyquist, H. L., Trans. Am. Soc. of Metals, 271, 1946.

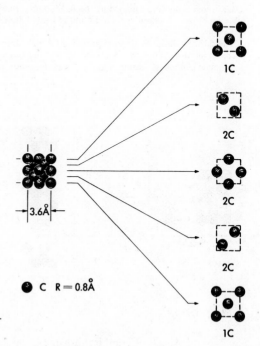

Fig. 5.7 Unit cell of diamond showing compactness.

The structure of the diamond is based on a cubic unit cell with eight atoms (Fig. 5.7). This structure, while characteristic of hard, brittle materials, does not completely explain the extreme hardness of this mineral. The diamond is an efficient abrasive, either as a powder or bonded into a wheel. Only its scarcity prevents it from taking over a large share of the abrasive industry.

11. Relative Abundance of the Elements

To summarize, the availability of ceramic materials depends on the degree to which nature has concentrated them and on the accessibility of the deposits. The abundance in the earth's crust (see Fig. 5.8) is not the important factor, for many commonly used elements such as mercury, silver, and antimony are less abundant than many of the so-called rarer elements.

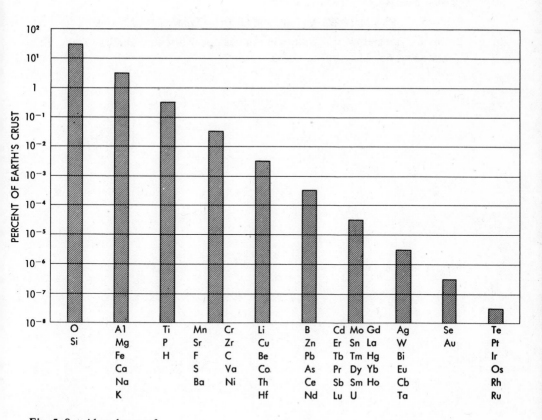

Fig. 5.8 Abundance of elements in the earth's crust.

References

Gruner, J. W., Crystal Structure of Talc and Pyrophyllite, *Proc. Nat. Acad. Sci.* **16**, 123, 1930

Page, B. N., Talc Deposits of Steatite Grade, Injo County, California, Cal. Div. of Mines, Special Report 8, 1951

Pask, J. A., and M. F. Warner, Fundamental Studies of Talc. I. Constitution of Talcs, *J. Am. Ceramic Soc.* **37**, 118, 1954

Rutten, M. G., Remarks on the Genesis of Flints, *Am. J. Sci.* **255**, 432, 1957

Heystek, H., and E. Planz, Mineralogy and Ceramic Properties of Some Californian Talcs, *Bull. Am. Ceramic Soc.* **43**, 555, 1964.

Sosman, R. B., The Phases of Silica, *Bull. Am. Ceramic Soc.* **43**, 213, 1964

Heffelfinger, R. E., *et al.*, Compositional Analysis of Twenty-six Zirconias, *J. Am. Ceramic Soc.* **47**, 646, 1964

Radford, C., Raw Materials. V. Flint and Quartz, *Ceramics* **16**, 22, 1965

Bruce, R. H., and W. H. Wilkinson, Fillers for Whiteware Bodies, *Trans. Brit. Ceramic Soc.* **65**, 233, 1966

6

Mining and treatment of the raw materials

1. Introduction

The mining and preliminary treatment of the raw materials are important steps in the production of ceramic ware, and therefore warrant a brief discussion here.

In the past, the mining of clays was often done on such a small tonnage basis that the methods employed were wasteful of labor compared with those used in the mining of metals. However, many mines are now using much more efficient methods, so that the cost of a ton of clay today is little more than it was a generation ago, in spite of a great increase in wage rates.

Great strides have also been made to improving the methods of treatment to give the customer a pure and uniform product. Examples of this are washed kaolins and ground feldspar. Also, as supplies of the higher grade materials are exhausted, more effort must be given to purification of lower grade deposits.

2. Mining

Open pit methods. The majority of clays are mined in open pits by stripping the overburden and taking out the useful clay. The stripping may be done by power shovels working on a face or, in some mines, by bulldozers and power scrapers of large capacity. The maximum economical depth of overburden depends on both the thickness and the value of the underlying clay layer.

The clay itself may be removed by power shovels, shale planers, or, if it is hard, by blasting. It is now commonly transported to the plant by truck rather than by rail. In some cases, such as in Cornwall, the clay is washed down into a sump by a powerful stream of water. Figure 6.1 shows sections of a number of typical clay-mining operations.

Harder materials like quartz rock, feldspar, and kyanite are quarried by the ordinary methods of drilling and blasting. Glass sands are often removed hydraulically.

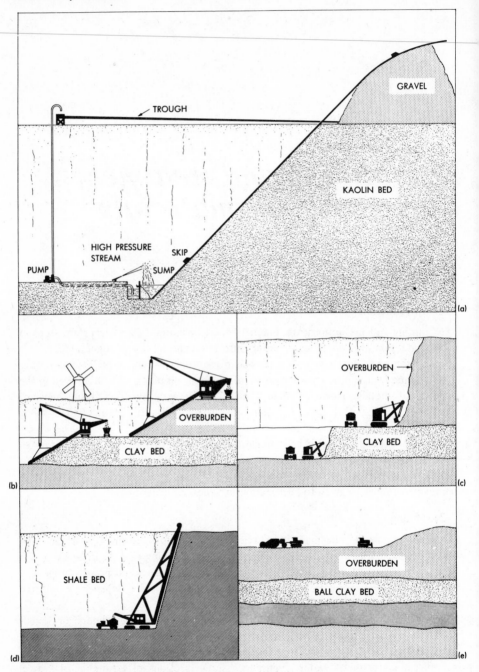

Fig. 6.1 Some typical mining operations in the ceramic industry: (a) English china clay mine in Cornwall; (b) removing fire clay in Germany; (c) removing Georgia kaolin with power shovels; (d) taking out hard clay with a shale planer; (e) stripping ball clay with earth-moving equipment.

Underground methods. Coal measure clays are usually taken from deep strata, using the conventional methods of sinking a shaft and then tunneling out into the proper level. As in all mining, a careful plan should be made, based on diamond drilling. Good timbering and clean, level tracks are essential for efficient operation.

3. Methods of Comminution

Theory. The reduction of a particle to two or more parts may take place in several ways, as shown in Fig. 6.2. In (a) the fracture takes place under simple compression if the material is brittle, in (b) by compression impact. In (c) the break is made by an impact too low in energy to fracture the whole piece but sufficient to remove a small corner. Breakage may occur as a result of one particle striking another at high velocity, as in (d). In some cases subdivision occurs by abrasion (e). In other cases subdivision occurs by abrasion (e). In the last case (f), a shredding action takes place in soft materials when a cutting tooth actually shaves off a fragment. In actual practice it is impossible to divide comminuting machines into rigid groups like those above, for in most cases several of the actions occur simultaneously.

In general, the energy required to break down a piece of material is proportional to the new surface area produced; thus the time and power expended increase rapidly as the size is reduced. The shape of the particles, as well as their distribution, varies both with the material and with the type of machine. Much study is still needed to give a complete picture of the comminution process.

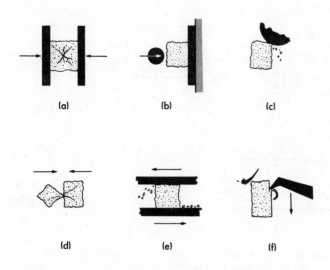

Fig. 6. 2   Principles of comminution: (a) compression; (b) compression impact; (c) nibbling; (d) self-impact; (e) abrasion; (f) shredding.

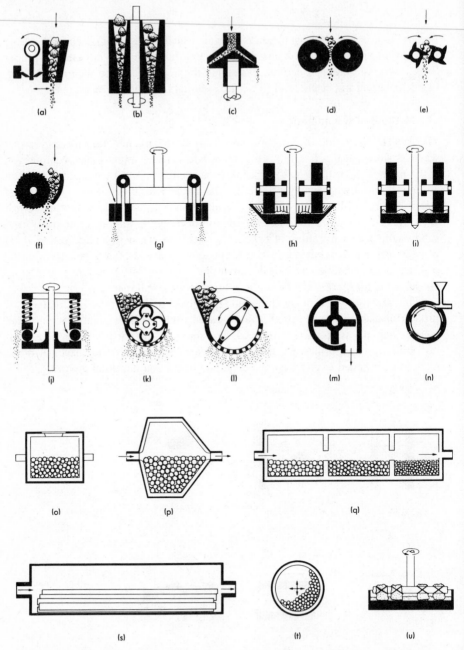

Fig. 6.3 Comminuting equipment: (a) jaw crusher; (b) gyratory crusher; (c) cone crusher; (d) smooth-roll crusher; (e) toothed-roll crusher; (f) single-roll crusher; (g) roller mill; (h) dry pan; (i) wet pan; (j) B and W mill; (k) ring roll crusher; (l) hammer mill; (m) dust blower; (n) fluid energy mill; (o) batch ball mill; (p) continuous ball mill; (q) tube mill; (s) rod mill; (t) vibrating ball mill; (u) rubbing mill.

Jaw crusher. As shown in Fig. 6.3(a), the jaw crusher is very much like a nutcracker, the fragments being crushed between a fixed and a moving jaw. These machines are important primary crushers, taking a feed of pieces up to a foot or more in diameter and delivering a product of pieces 1 to 3 inches in diameter. The capacity may be as high as 100 tons per hour. The feed may be hard or medium-hard material.

Gyratory crusher. This machine, shown at (b), has the same action as the jaw crusher, since the inner cone oscillates with a circular motion but does not revolve. It is used for brittle materials like magnesite and limestone, and has a capacity of up to several hundred tons per hour.

Cone crusher. This machine, shown at (c), operates on the same principle as the previous one, except that the cone is more obtuse to give a greater discharge area. It is used a great deal for feldspar grinding. It brings the material down to a 20-mesh size, and it has a capacity of up to 10 tons per hour.

Roll crusher. The roll crusher (d) is used for crushing grog and other brittle materials from 1-inch size down to 8 mesh or even finer. The squeezing action is continuous, and capacities up to 10 tons per hour are common.

Toothed rolls. In this case (e), the rolls are corrugated or toothed and one often turns faster than the other. Since they do not tend to clog, these rolls are used for soft materials such as lump clay. The capacity is great, as much as 180 tons per hour.

Single-roll crusher. This mill, shown at (f), has a high capacity and is used for medium-soft materials like limestone.

Roller mill. As shown at (g), this is a roll crusher in which the grinding pressure is exerted by centrifugal force. It is used for medium-hard and soft materials and can produce 325-mesh material in some cases.

Dry pan. This machine (h) is extensively used in the ceramic industry for grinding ganister, grog, shale, or flint clay. There are many modifications, but in general the charge is fed under the mullers with plows, and the fines sift out through the screen plates in a continuous process. They are made in several sizes; a large machine might be 10 feet in diameter and grind 50 tons per hour.

Wet pan. This is similar to the previous machine, but is used for batch mixing and grinding of wet mixes (i).

Ball pulverizer. This mill (j) is like a huge ball bearing grinding between the balls and races. It is used mainly for pulverized coal, but is also satisfactory for soft materials like bauxite.

Ring roll crusher. This mill, shown in (k), is similar to the preceding one except for the design of the hammers.

Hammer mill. This mill (l) is used for brittle materials. A series of hammers continually strikes the feed, reducing it until it can pass out through the screen plate. Capacities up to 150 tons per hour can be reached.

Disintegrator (dust blower). This machine (m) is a heavy centrifugal fan used for breaking up lump clay or filter cakes in preparation for dust pressing.

Steam pulverizer. This mill (n) is unique in that there are no moving parts. The comminution takes place as a result of the impact and abrasion of the particles on each other and on the lining. It is capable of fine grinding, since there is a classifying action. However, the steam consumption per ton is quite large.

Ball mills. These mills are used extensively in the ceramic industry for finely grinding such materials as quartz, feldspar, and cement clinker. They may be used either wet or dry. The simplest form (o) is a hollow cylinder lined with stone or porcelain and containing hard balls. As the cylinder revolves, the balls tumble over one another and grind the material between them. The grinding efficiency depends on many factors, such as rate of turning, size of the balls, specific gravity of the balls, and amount of charge.

Another type of mill may be used continuously, as shown in (p). The conical shape segregates the balls of different sizes for efficient grinding. Still another type known as a tube mill (q) is used for continuous milling.

The rod mill (s) employs rods rather than balls. It is widely used for ores, but only to a small extent for ceramics because of the iron contamination. A new type of mill (t) vibrates the mass of balls at high frequency to achieve very rapid fine grinding.

Ball mills regularly grind below 200 or 325 mesh and sizes averaging as low as 5 microns may be obtained. A typical mill for feldspar might have a cylinder 7½ ft in diameter and 10 ft long inside. It would require 85 hp, take ¾-in. feed, and deliver 1 ton per hour of 90 percent of the material through a 325-mesh screen.

Rubbing mills. This early type of mill is shown in (u). Heavy stones are dragged over a stone base and abrade the wet mix. Such mills have a low power efficiency and are little used in this country.

4. Size Classification

Screens. Size classification is carried out by means of screens down to 120 mesh, or in some cases to 325 mesh. These screens are usually woven from bronze wire; they form a series shown in Table 6.1. Coarser screens are often made from perforated plates or bars as shown in Fig. 6.4. The taper illustrated here does much to keep the screen from clogging. Recently a considerable increase in the efficiency of any screen has been made possible through heating the screen wires by electrical conduction. This prevents clogging and increases the life of the screen. Silk screens are often used for fine abrasive materials, since they last longer than metal screens.

Table 6.1 Tyler series screen

Openings, inches	Openings, millimeters	Mesh per lineal inch	Diameter of wire, inches
1.050	26.67		0.148
0.883	22.43		0.135
0.742	18.85		0.135
0.624	15.85		0.120
0.525	13.33		0.105
0.441	11.20		0.105
0.371	9.423		0.092
0.312	7.925	$2\frac{1}{2}$	0.088
0.263	6.680	3	0.070
0.221	5.613	$3\frac{1}{2}$	0.065
0.185	4.699	4	0.065
0.156	3.962	5	0.044
0.131	3.327	6	0.036
0.110	2.794	7	0.0328
0.093	2.362	8	0.0320
0.078	1.981	9	0.0330
0.065	1.651	10	0.0350
0.055	1.397	12	0.0280
0.046	1.168	14	0.0250
0.0390	0.991	16	0.0235
0.0328	0.833	20	0.0172
0.0276	0.701	24	0.0141
0.0232	0.589	28	0.0125
0.0195	0.495	32	0.0118
0.0164	0.417	35	0.0122
0.0138	0.351	42	0.0100
0.0116	0.295	48	0.0092
0.0097	0.246	60	0.0070
0.0082	0.208	65	0.0072
0.0069	0.175	80	0.0056
0.0058	0.147	100	0.0042
0.0049	0.124	115	0.0038
0.0041	0.104	150	0.0026
0.0035	0.088	170	0.0024
0.0029	0.074	200	0.0021
0.0024	0.061	250	0.0016
0.0021	0.053	270	0.0016
0.0017	0.043	325	0.0014
0.0015	0.038	400	0.0010

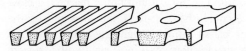

Fig. 6.4   Heavy screen plates with nonblinding openings.

In order to ensure a constant flow of material through the screen, the mesh must be vibrated in some way.

Screening may be done wet or dry, but for the very fine sizes wet screening is generally more efficient. Several screens are often superimposed, with the coarse ones at the top, to separate the feed into several classes and to prevent overloading the finer screens.

Air classifiers. When the final product is used in the dry condition it is generally more economical to dry grind to avoid the expense of final drying. Therefore, such products as flint and feldspar are dry ground. During the grinding operation, if it is continuous, the fines are swept out of the mill by a current of air or carried away by a mechanical conveyor. They are put through a classifier which takes out the coarse material for return to the mill; then the fines go into storage.

The rate of settling of fine particles in air or water is given by Stokes' law as follows:

$$v = \frac{2}{9} \frac{g(\rho_1 - \rho_2)}{\eta} r^2 ,$$

assuming spheres or equivalent spheres. In this equation, ρ_1 is the density of the solid particles and ρ_2 that of the fluid, r is the radius of the particle, and η is the viscosity of the fluid. Values of v obtained by using the appropriate constants are shown in Table 6.2.

Table 6.2 Settling velocity of particles in air and water

Particle size in equivalent spherical diameter in microns	Settling velocity in cm per second	
	In air	In water
1	0.0077	0.000082
2	0.031	0.00032
5	0.19	0.0020
10	0.77	0.0081
25	4.8	0.050
44 (325 mesh)	15	0.16
74 (200 mesh)	42	0.44
104 (150 mesh)	81	0.87

If a dust-laden column of air or water is rising vertically in a cylinder, those particles having a settling velocity less than the fluid will rise to the top and those with a greater value will stay at the bottom. After equilibrium is reached, a close division into two size fractions is accomplished. This is the principle of the elutriator, an important machine for classification. Often the classification is hastened by centrifugal action in the air classifier. A typical air classifier system connected with a ball mill is shown in Fig. 6.5(a).

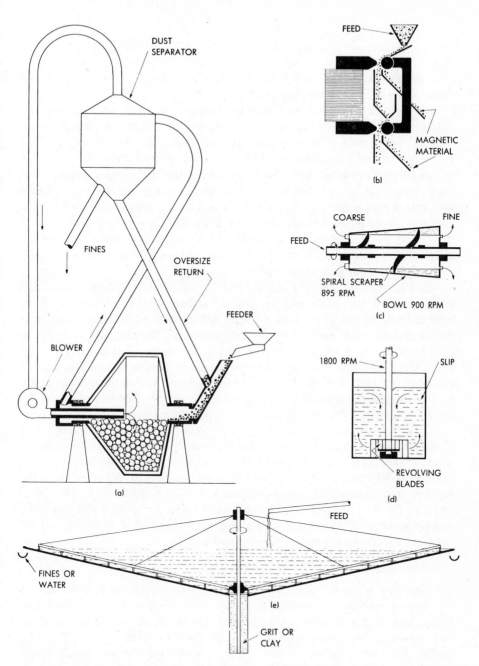

Fig. 6.5 Classification operations: (a) continuous ball mill with air separator; (b) magnetic separator; (c) dewatering centrifuge; (d) high-speed blunger; (e) Dorr-type bowl classifier.

Water classification. This method is similar to air classification except that the supporting medium is water. It is used extensively for taking the coarse material from kaolin. In Europe much of the flint and feldspar is wet ground and continuously classified by settling in water.

In England, china clay is washed by passing a slip of about 2 percent solid concentration at a velocity of 1 to 2 feet per second through long troughs called "micas" that contain riffles where the grit and mica settle out. The clay is then settled in large tanks and dried out or filter-pressed to remove the water.

Another washing method consists of passing the well-blunged and deflocculated slip through a bowl classifier, as shown in Fig. 6.5(e). This is a large, shallow tank with a conical bottom. As the grit settles, a slow-moving scraper pushes it out the center hole in the bottom, while the clean slip passes over the edge into a launder where it is flocculated and fed to a similar but larger classifier for dewatering. A classifier 25 feet in diameter with a slip of 2 percent solids will degrit 50 tons of clay a day with about 3 percent clay loss.

Still another classifier for washing clay is the continuous centrifuge. Figure 6.5(c) shows a cross section of this machine. The grit is thrown to the inner surface of the bowl and scraped out of the small-diameter end of the bowl, while the clean slip is thrown out of small holes at the large-diameter end. A centrifuge 36 inches in diameter driven by a 15 hp motor at 1400 rpm will degrit 10 tons of dry clay per hour. The clean slip may then be dewatered by passing it through another centrifuge running at higher speeds. Well-engineered gearing is needed in this machine to give the slow differential speed to the scraper. The flow sheets at the end of the chapter show some of these processes (Figs. 6.10 to 6.15).

The hydrocyclone, a relatively simple piece of equipment, is now used extensively for classifying fine suspended particles. It consists of a hollow cone set on a vertical axis with the small end downward. The suspension is forced tangentially into the cone near the top, setting up a high-speed rotary motion of the liquid so that classification can be continuous. A cross section of such a classifier is shown in Fig. 6.6. A 6-inch diameter cone has a capacity of around 100 gpm, using an inlet pressure of 33 psi and with a separation range of 30 to 40 microns. Of course, smaller cones will make a finer separation.

5. Disintegration

It is necessary to break down clays to their ultimate particles as a preliminary to any washing process, or sometimes to prepare a slip from washed clay filter cakes. This is done in a wet mixer called a blunger. The old types were large cylindrical tanks in which slowly moving paddles were mounted on a vertical shaft. The more modern types use a high-speed motor that circulates the slip and rapidly breaks down lumps, as shown in Fig. 6.5(d).

6. Chemical Treatments

Ceramic raw materials, except for chemicals used in glazes or special refractories, are seldom chemically treated because of the added cost. An exception is sea-water

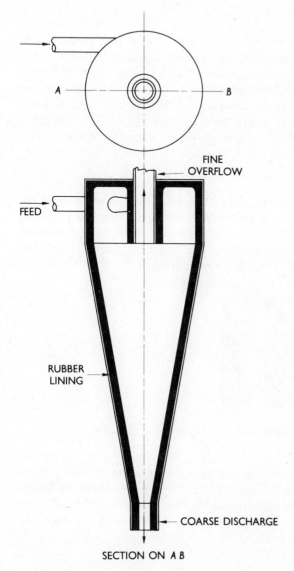

A

B

FINE
OVERFLOW

FEED

RUBBER
LINING

COARSE DISCHARGE

SECTION ON *A B*

Fig. 6. 6   Cross section of a hydrocyclone classifier.

magnesite, which is now produced in large quantities by the method shown in the flow sheet of Fig. 6.13.

In Europe, glass sands are sometimes treated to remove iron, while here some paper clays are bleached with zinc hyposulfite.

The materials for some of the special refractories are produced chemically, for example, alumina, zirconia, and beryllia, but since these processes serve the ceramic industry to only a small extent they will not be described here.

7. Magnetic Separation

This method is widely used, especially to remove iron or iron minerals from feldspar. A cross section of a high-intensity magnetic separator is shown in Fig. 6.5(b). Minerals with rather low susceptibilities may be taken out in the intense field generated. A relatively small unit will handle 2 to 4 tons of feldspar per hour with a power consumption of only 1 kw-hr per ton. It is, of course, necessary to grind the mineral fine enough to unlock the magnetic grains. On the other hand, very finely ground material does not pass through a magnetic separator easily because of caking and uneven flow.

8. Froth Flotation

In the last twenty-five years the methods of flotation have revolutionized the treatment of ore minerals. This method consists of mixing the finely ground, water-suspended ore with a frothing agent, whereby there is a differential adsorption by the bubbles on the ore particles and the gangue particles. One or the other floats to the surface and is removed, as shown in Fig. 6.7.

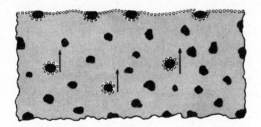

Fig. 6. 7   Froth flotation due to selective adsorption of bubbles on one of the minerals in suspension.

In the ceramic field, flotation has not been used to a great extent in actual production, although considerable research has been conducted with various minerals. At present the feldspar industry uses flotation to some extent in removing quartz from feldspar. Also, mica flakes are floated from sand and quartz as a kaolin by-product.

9. Filtering

Water removal by filtering is a common practice in the ceramic industry, since it tends to take out soluble salts. The plate-type filter press is generally used (Fig. 6.8).

10. Drying

Bulk drying is still carried out in open sheds under natural conditions, but in modern plants it is done in rotary dryers, generally of the indirect type shown in Fig. 6.9, to prevent contamination by combustion products.

Fig. 6. 8 Filter press being unloaded. (Josiah Wedgwood and Sons, Ltd.)

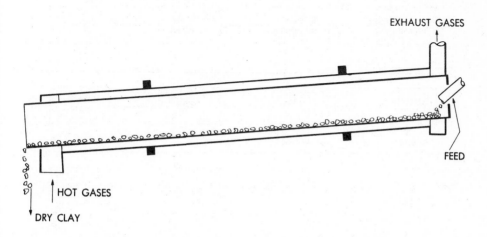

Fig. 6. 9 A cross section of an indirectly heated rotary dryer for treating lump clay.

Some lump material, such as filter cakes, may be dried on cars in humidity controlled continuous driers or on conveyor-type driers.

Spray drying of clay slip, done by injecting it into a hot chamber, has been used for some clays, but a very careful temperature control is needed to avoid breaking down the clay structure and thus reducing subsequent plasticity.

11. Storage and Handling

Ceramic industries, such as a whiteware pottery, that use relatively small amounts of raw materials can afford to store a supply sufficient for weeks or even months. In these cases material is shipped from some distance and an interruption of flow must be guarded against. On the other hand, refractory and heavy clay products manufacturers usually mine their clay close at hand and use such large quantities that a day's storage may be all that is economical.

In the European potteries storage serves another purpose, that of aging. In many plants the clays are left outdoors for a year or more to freeze and thaw or to dry and be wetted by rains, processes which undoubtedly improve their working properties.

The trend in handling these raw materials is toward substituting conveyors for hand labor as far as possible. There is not space here to go into the many types available, but a visit to a modern plant will show that many labor-saving devices are used.

Flow sheets. A number of flow sheets for material processing are shown in Figs. 6.10 to 6.15.

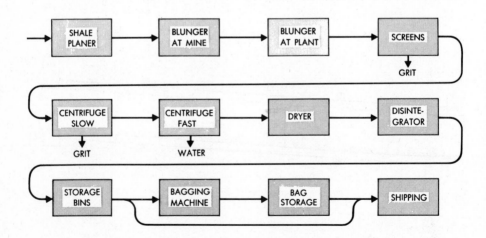

Fig. 6.10 Flow sheet illustrating the preparation of Georgia kaolin.

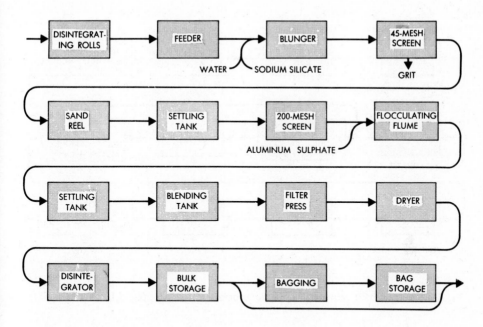

Fig. 6.11   Flow sheet of European method of producing washed kaolin.

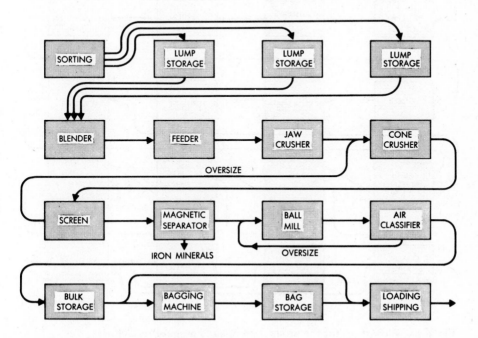

Fig. 6.12   Flow sheets of feldspar milling.

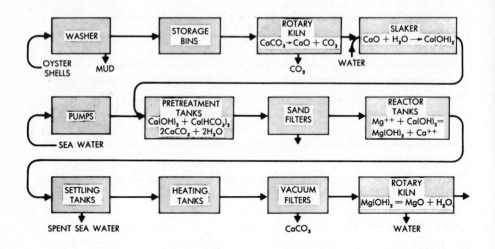

Fig. 6.13 Flow sheet showing the lime process of taking magnesia from sea water.

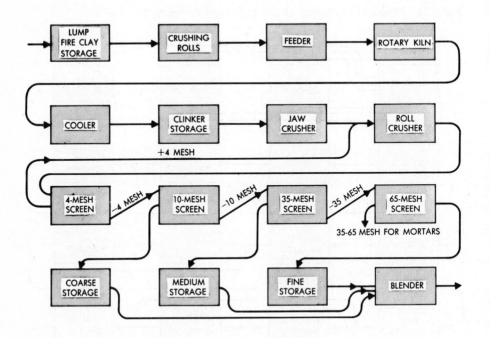

Fig. 6.14 Flow sheet for the production of sized grog for refractories.

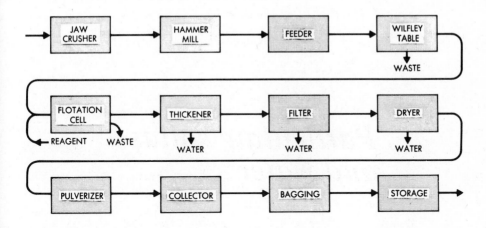

Fig. 6.15   Flow sheet for the treatment of talc by froth flotation.

References

Powell, H. E., and W. A. Calhoun, *The Hydrocyclone in Clay Beneficiation*, U.S. Bureau of Mines. R. I. 6275, 1962.

Lissenden, A., Vibratory Grinding-mill, *Chem. Proc. Eng.* **46**, 203, 1965

Rockwell, P. M., and A. J. Gitter, Fluid Energy Grinding, *Bull. Am. Ceramic Soc.* **44**, 497, 1965

Fraas, F., *Magnetic Separation of Minerals of Low Susceptibility and Small Particle Size*, U.S. Bureau of Mines, R.I. 7292, 1969

7

Particulate solids and water

1. Introduction

Since the great majority of ceramic forming processes depend on the particle-water system, it seems worthwhile to devote a chapter to it. At present we have by no means a clear-cut picture of this system, but the main features can be brought out.

2. Elements of Colloidal Chemistry

A colloid is any type of insoluble material in an aqueous suspension having a particle size so small that surface effects overshadow bulk effects. The upper diameter of colloidal particles is around 5 microns. These particles may be lyophobic (shunning water) or lyocratic (attracting water). Nearly all ceramic materials belong to the latter class.

Each suspended particle is surrounded by a hull of water. The closely bound water molecules build up a thickness of about 15 Å, whereas loosely held molecules have a thickess of perhaps 100 Å more. These water molecules are polarized and thus have their positive ends attracted to the negatively charged clay mineral surface. In addition to this the clay minerals have monovalent ions adsorbed on the surface (Fig. 7.1) and these are hydrated. Some authors believe the adsorbed water has a structure conforming with the atomic structure of the clay mineral surface.

The clay mineral particle is a very specialized colloid because it has two distinct surface areas, the faces and the edge. It is quite possible, therefore, to have negative charges on the edges and positive ones on the faces. This makes possible the bulky, "house of cards" floc shown in Fig. 7.2. On the other hand, in a deflocculated condition there must be negative charges all over the surface to cause a plate-to-plate repulsion.

In addition to deflocculation by means of electrolytes (pH), it may be accomplished by surface-active ions such as hexadecyl trimethyl ammonia. The polyphosphates, sulfonic acid derivatives, and tannates probably belong to this class of deflocculants.

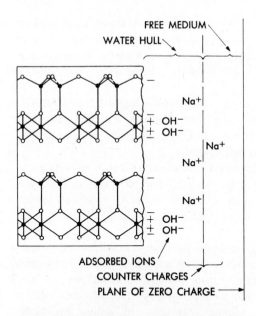

FREE MEDIUM
WATER HULL

Na⁺
OH⁻
OH⁻
Na⁺
Na⁺
Na⁺
OH⁻
OH⁻

ADSORBED IONS
COUNTER CHARGES
PLANE OF ZERO CHARGE

Fig. 7. 1 Broken edge of a kaolinite crystal showing lyosphere.

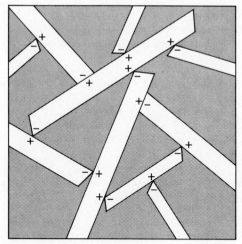

Fig. 7. 2 "House of cards" structure in a kaolinite floc.

3. Deflocculation of Slips

As in all precise work with clay, it is advisable to carry out the first experimentation on clean, monodisperse fractions of kaolinite. Figure 7.3 shows the curve of apparent viscosity of a clean Florida kaolin suspension of 16 percent solids as the NaOH content is increased. At the origin the clay is free of all adsorbed ions except H^+ and $(OH)^-$, but as NaOH increases these hydrogen ions are gradually replaced by Na^+. There is no marked change in viscosity until the base exchange capacity is reached at 3.7 milliequivalents, where all the adsorbed hydrogen ions are replaced. As soon as there is any excess NaOH in the suspension, the viscosity drops suddenly to only 1/200 of its original value, a remarkable change.

On the same plot is shown a curve of hydrogen ion concentration or pH. This value shows a sudden rise to the basic condition at exactly the same content of NaOH that is required to drop the viscosity. In other words, the suspension must have free Na^+ and OH^- to bring about deflocculation.

A fairly complete survey of deflocculants has shown that they must have two characteristics, one a basic reaction and the other a monovalent cation. Therefore, all deflocculants are salts of the alkali metals or ammonia, such as sodium carbonate, sodium silicate, and sodium hydroxide, that hydrolyze to give a basic reaction.

In Fig. 7.3 it was shown that sodium hydroxide was an excellent defloctulant for a clean clay. However, practice has shown that it is not satisfactory for use with commercial clay slips, whereas sodium carbonate and sodium silicate are excellent. This may be explained by the fact that commercial clays have adsorbed calcium ions which form the slightly soluble calcium hydroxide with sodium hydroxide. Thus divalent ions are produced in solution, a fact that hinders deflocculation. On the other hand, both sodium carbonate and sodium silicate form the relatively insoluble calcium carbonate or silicate and thus remove the Ca^{++} from solution.

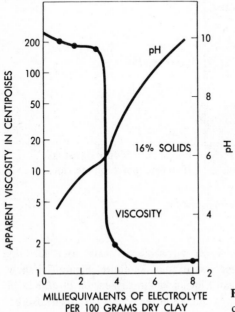

Fig. 7. 3 Viscosity and pH of a clean, monodisperse fraction of kaolinite.

Theory of deflocculation. If a suspension of monodisperse hydrogen clay particles is allowed to settle for a long time, the particles will collect in a layer at the bottom of the container. When this equilibrium condition is reached, there will be a balance between repulsion forces, attractive forces, and gravity forces.

If the settled clay layer has the water decanted off the top and is dried very slowly, it is possible to measure the volume and calculate the water film thickness as previously explained. This thickness is found to be greater than that in plastic clay, since the forces are less. However, the bulk density of the dried piece will be lower than that of a corresponding piece dried from a plastic mass since there has not been an opportunity for the particles to pack closely.

Furthermore, the suspension discussed above, when examined under the microscope, is seen to be composed not of individual particles, but of small flocks of many particles. For this reason the suspension settles rapidly and leaves a sediment of low density.

Contrast this with a similar kaolinite fraction that has been deflocculated with NaOH. Here the particles are individuals that move about by Brownian movement; therefore, the settling is very slow. If sufficient time is given, however, the sediment is more dense than for the previous case, since each crystal fits into a closely packed arrangement. This at times produces an almost rocklike sediment.

From this simple evidence it may be deduced that in the flocculated system there are between the particles attractive forces that draw them together into flocks; however, at a certain distance the repulsion forces increase to balance them and maintain a stable condition. On the other hand, in deflocculated suspensions there is no evidence of any attractive force. This is confirmed by the fact that flocculated systems have definite yield points but deflocculated systems do not.

This force system is represented with a reasonable degree of certainty in the diagram of Fig. 7.4. Here are plotted the repulsion forces between particles, which are quite precisely known, and the attractive and total forces *a*, which are only estimated. It will be noted that the force between particles is zero at a spacing (water film) of 0.04 micron. This is a point of equilibrium because the slope of the total force line passing through the axis is negative. That is, if the particles are slightly separated beyond this distance, the forces built up are attraction forces and tend to bring them together again. On the other hand, if the particles are forced together, repulsion forces are built up to push them apart. This is a simple, clear-cut picture and seems to explain the behavior of clay suspensions.

In the case of a deflocculated suspension, the total force curve is represented by *b*. Since there are no attractive forces, there can be no yield point. Curve *c* is the total force for a flocculated system at higher pressure (curve *a* moved upward) whereby the stability point is moved to the left. This picture is complicated by the possibility of polarization of the clay plates, that is, there being some areas with positive charges and some with negative charges. This condition would aid in forming scaffoldlike flocks.

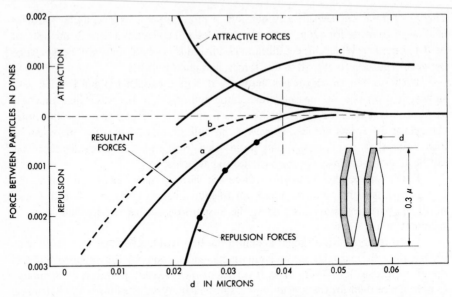

Fig. 7. 4 The forces between a pair of kaolinite plates in suspension.

Mechanism of deflocculation. We have discussed what happens in deflocculation, but to explain why it happens is not at all simple. The theory most generally accepted by the colloid chemists is briefly described below, but it should be remembered that at present it is only a theory and there is little proof to substantiate it. Figure 7.1 showed a section of the kaolinite structure with an edge having broken bonds. It is supposed that $(OH)^-$ are preferentially absorbed at the positive points in the lattice structure. In the hydrogen clay H^+ act as counter ions in the surrounding water hull. The size of this hull changes with the type of counter ions, becoming larger with the more highly hydrated Na^+. As the NaOH concentration increases, the negative charge on the particle increases, because the increased Na^+ outside the water hull will leave some $(OH)^-$ unneutralized. The hydrogen clay should have no charge and a pH of 7, but this is not exactly realized because of dissolved impurities in the distilled water. Adding NaOH up to the base exchange capacity makes no great change in pH because of a sort of buffer action, but as soon as the base exchange capacity is exceeded the Na^+ and $(OH)^-$ will exist in the free water medium and will prevent the building up of any attractive forces.

4. Laws of Flow

A vital characteristic of clays and of most ceramic bodies is the nebulous quality known as plasticity. This is the quality that permits the forming of the piece, often in very intricate shapes, and then keeps the forces of gravity or the shocks inherent in manufacture from deforming it.

There are several types of flow that are well understood, such as viscous and elastic flow; in addition, there are a myriad of substances that do not see fit to follow these simple laws, but exhibit a flow of considerable complexity. Indeed, it is sometimes hard to distinguish between a liquid and a solid. While a perfect liquid is defined as a medium incapable of supporting shear, there are suspensions that under some conditions have a yield point and under others do not. Among these are clay suspensions. An attempt will be made first to explain the simple flow types and then to take up the more complicated ones.

Viscous state. Viscous flow is found in homogeneous liquids moving at low velocities. The flow may be expressed simply by

$$F = k \frac{dv}{dr}, \tag{1}$$

where k is proportional to viscosity. In other words, the flow rate is proportional directly to the force applied and inversely to the viscosity. Consider the 1-cm cube shown in Fig. 7.5. If a force of 1 dyne tends to shear the cube at a velocity of 1 cm per sec, then the viscosity of a liquid making up the cube is unity. The dimensions of viscosity are given by

$$\eta = \frac{F}{\dfrac{dv}{dr}} = \frac{\text{dyne} \cdot \text{cm}^{-2} \cdot \text{cm}}{\text{cm} \cdot \text{sec}^{-1}}$$

$$= \text{dynes} \cdot \text{cm}^{-2} \cdot \text{sec}, \tag{2}$$

or in dimensions of mass, length, and time,

$$MLT^{-2}L^{-2}T = ML^{-1}T^{-1} \tag{3}$$

When considering any physical property measured quantitatively, it is always well to determine the units of that quantity by dimensional analysis so that both sides of an equation will be dimensionally compatible. An excellent short treatment of this subject may be found in Mark's *Mechanical Engineers' Handbook*.

The unit of viscosity is the *poise* and it is well to keep in mind that water at room temperature has a viscosity of close to 0.01 poise, or one centipoise.

If more and more shearing force is applied to a fluid, the velocity of flow will increase to a point where the viscous or laminar flow breaks down and turbulence sets in. In the turbulent region another law of flow is followed, important in hydraulics and aerodynamics but not in ceramics.

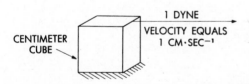

CENTIMETER
CUBE

1 DYNE
VELOCITY EQUALS
1 CM·SEC⁻¹

Fig. 7. 5 Cube of fluid undergoing shear.

Elastic state. A perfectly elastic solid is one which obeys Hook's Law — that is, the deformation is proportional to the deforming force up to the breaking point. When the deforming force is removed, the solid resumes its original form. There are, perhaps, no perfectly elastic materials in nature, but many solids approach this condition so closely that they may be considered as such for all intents and purposes.

Plastic state. Unfortunately many materials, including most ceramic materials, do not have the simple flow properties described for fluids and elastic solids, but instead reveal a complexity that makes the subject very difficult to analyze. The study of flow in such materials as rubber, fats, waxes, paints, ceramic pastes, clay slips, and even metals is called rheology.

The various types of flow are summarized clearly by Scott Blair (see References). The types that interest us in ceramics are the visco-elastic (glass) and plastico-elastic (clay pastes). The former will be considered in Chapter 12; the latter will be treated here in some detail.

When a piece of clay paste is stressed with an increasing force, a stress-strain diagram like that in Fig. 7.6 will be obtained. Up to the yield point a, the flow is elastic. Should the stress be released after a very short time interval, the original size would be regained. However, should the stress be maintained for a long time, some of this ability to return would be lost, probably by migration of water from the highly stressed parts to those with lower stress. Carrying the stress beyond the yield point produces plastic flow and allows a considerable deformation of the piece before cracking appears at b. There are, then, two features of importance in this diagram; first, the yield value and, second, the extension at breaking.

A workable clay paste should, therefore, have a yield value high enough to prevent accidental deforming and an extension large enough to allow forming without fracture. For a given clay paste these two features are not independent of each other, for by varying the water content it is possible to increase either one, while at the same time the other decreases, as is shown in the stress-strain diagrams

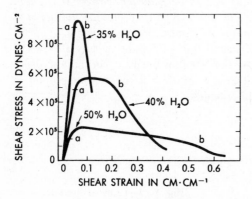

Fig. 7. 6 Stress-strain diagram for a plastic clay.

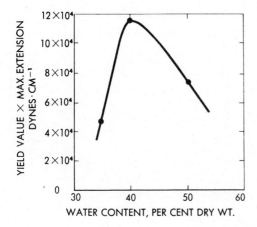

Fig. 7. 7 Workability of a clay paste at different water contents.

of Fig. 7.6. We may then say that the workability can be approximately evaluated by the product of yield point and maximum extension, and that a given water content produces a maximum in this value as shown in Fig. 7.7.

In general, all plastico-elastic materials consist of at least two phases, solid and fluid; for example, clay pastes, paints, plasticene, and even metals have hard crystals and soft intergrain material.

Flow of dispersed suspensions. The flow of clay-water suspensions is of particular interest to ceramists, so some space will be devoted to its consideration. When such a suspension is sheared at a uniform rate and the shearing force measured, a value of consistency which we may call apparent viscosity results from Eq. (2). This is a convenient term to use.

It has been found that the important variable in the dispersed clay-water suspension is the concentration of solids. The other variables such as the sizes and shapes of the particles have less effect.

As the suspension is sheared the disclike clay particles revolve about an axis in the plane of shear (Fig. 7.8). This rotation absorbs energy in itself, but should the particles approach each other because the suspension is more concentrated the mutual interference will cause further energy absorption.

It may then be assumed that the total apparent viscosity of the suspension will be due to three contributing factors; first, the energy absorbed in the water itself; second, the energy absorbed by the individual particle; and third, the energy of the particle collisions. Then we may write

$$\eta_s = \eta_l(1 - C) + kC^n + k_1 C^m, \tag{4}$$

in which n is found by experiment to be unity, m is 3, and the values of k and k_1 are 0.08 and 7.5 respectively for dispersed suspensions. To enable the reader to visualize this suspension, a section of it has been drawn to scale in Fig. 7.9 for a

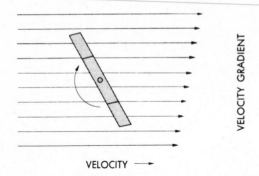

VELOCITY GRADIENT

VELOCITY ⟶

Fig. 7. 8 A plate-shaped particle revolving in a fluid undergoing shear.

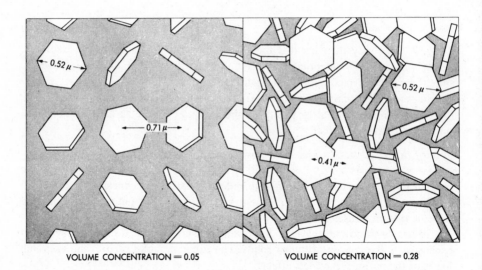

VOLUME CONCENTRATION = 0.05   VOLUME CONCENTRATION = 0.28

Fig. 7. 9 A cross section of a suspension of monodisperse kaolinite particles at two concentrations.

volume concentration C of 0.05 and 0.28. It will be seen that in the dilute suspension each particle has room to turn with little chance of interfering with its neighbor. When C increases to a point where the center-to-center distance reaches the diameter of the disc, as in the more concentrated suspension, then the viscosity increases with great rapidity. In other words, the first term of Eq. (4) is the important one for very dilute suspensions, the second term for medium concentrations, and the third term for high concentrations.

In the previous discussion the shear rate was considered a constant, but should this value be varied with everything else constant, it would be expected that η_s would be invariable as in the case of Newtonian fluids. This is not the case, however, for suspensions of nonisotropic particles such as clay. There is some evidence that suspensions of spherical particles behave like fluids, but the story is by no means complete.

In Fig. 7.10 are shown curves of η_s for various rates of reciprocal shear, $1/(dv/dr)$. It will be seen that the apparent viscosity decreases with an increasing shearing rate. If the curve is extended to zero in the plot (infinite shear rate), a value is obtained called the basic apparent viscosity, which is a convenient value to use for all suspensions.

The property of decreasing apparent viscosity by increasing the shear rate is called thixotropy. A quantitative measure suggested by Goodeve (1939) is the slopes of the curves of Fig. 7.10 at the origin of the plot, or at infinite shear. The coefficient of thixotropy of suspensions may be expressed by

$$\theta = k_3 C + k_4 C^3, \tag{5}$$

where k_3 and k_4 are constants depending on the particle size and shape and on the degree of flocculation. Some authors consider thixotropy to be a change of apparent viscosity with time at constant shear rate. It is interesting to note that the unit of reciprocal shear rate is seconds. The units of θ are dynes $\cdot$ cm^{-2} or $ML^{-1}T^{-2}$.

The cause of thixotropic behavior is by no means clear at present. One theory, for example, assumes that a scaffold-type structure is set up when the particles are at rest and that this is gradually broken down as the shear rate increases. This does not seem plausible for deflocculated systems in which the particles are individuals. This subject needs much more study, since thixotropy is of great importance in the industrial use of casting slips.

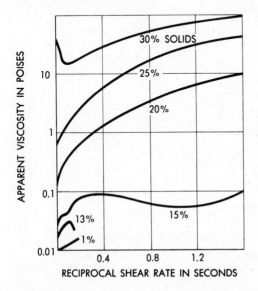

Fig. 7.10 Influence of shear rate on apparent viscosity.

Another property of slips and pastes sometimes met is that of stiffening when agitated. This is called dilatancy and is encountered with coarse clays and sands. It is believed to be caused by a change from close packing, where there is sufficient water for lubrication, to a more open packing where there is too small a volume of water to fill the voids.

Rheopexy is a term used to describe the gelation produced in a colloid by gentle agitation.

Another property of suspensions is the yield point, or the maximum shearing strain that may be supported before continuous shear starts. It has not been possible with our most sensitive instruments to detect any yield point in deflocculated suspensions.

Flow of flocced suspensions. A flocced suspension has a much higher coefficient of apparent viscosity than the same suspension when deflocculated. In the former there are groups of particles rather than single ones, and these groups offer great resistance to shear. However, the apparent coefficient of viscosity of the flocced suspension may be expressed by Eq. (4) with different values for k and k_1. Figure 7.11 gives curves of η_s for a series of monodisperse fractions, to show the effect of particle size. In the same way the thixotropy for the flocced suspensions may be expressed by Eq. (5), but with different values of k_2 and k_3.

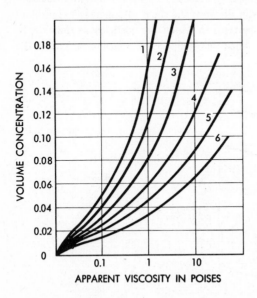

Fig. 7.11 The apparent viscosity of several monodisperse fractions of kaolin in aqueous suspension (flocced). Figures refer to equivalent spherical diameter in microns: (1) 6.4 to 12.8; (2) 3.2 to 6.4; (3) 1.6 to 3.2; (4) 0.8 to 1.6; (5) 0.4 to 0.8; (6) 0.2 to 0.4.

A yield point is always present in a flocced suspension, even with a solid concentration of only 0.1 percent by volume. It may be expressed by

$$F_0 = k_4 C^3, \tag{6}$$

where k_4 is a constant depending on particle shape and on size and degree of flocculation. The units are dynes·cm^{-2}, or the same as for thixotropy.

Figure 7.12 shows some values of yield point for a series of flocced, monodisperse kaolinite fractions. The cause of the yield point is the action of elastic forces between the particles.

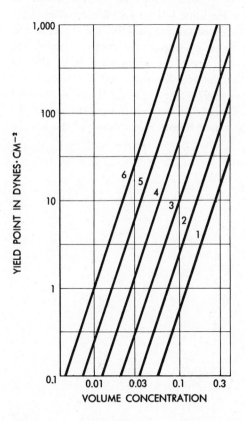

Fig. 7.12 Yield point of flocced suspensions. (Particle size as given in Fig. 7.11.)

Influence of particle shape on flow of suspensions. A study made with similar fractions of various particles of increasing elongation indicates that the more isometric the particle in deflocculated suspensions, the lower the coefficient of apparent viscosity. In other words, it takes more energy to rotate elongated particles than spherical ones. However, when the system is flocced, the particle shape has no influence, as clumps rather than individual particles are encountered.

5. Mechanism of Plasticity

Introduction. The subject of plasticity has been discussed from many points of view, but there is not space here to review the many theories that have been devised. The reader is again referred to Scott-Blair for a summary. In this section there will be set forth the facts as we know them, and then the theory will be described that seems the most plausible to the author. However, much more study of this problem is needed before the picture can be made clear.

Difference between suspensions and plastic masses. If the yield point of a monodispersed, flocced suspension is plotted over the widest range of concentration possible for measurement, the same is done for plastic masses of the same particles, and the values are then placed on a single plot, the curves in Fig. 7.13 are obtained. In the first place there is a gap in the sticky range of clay consistency in which there is at present no way of measuring yield point; this is a range which cannot be used in ceramic production. Second, it will be observed that the slopes for the two consistency regions are different. For the slips it is 3 and for the solids 6, which indicates that a different law governs the consistency of the solid; otherwise the points for the solid would follow an extension of the line for the suspensions.

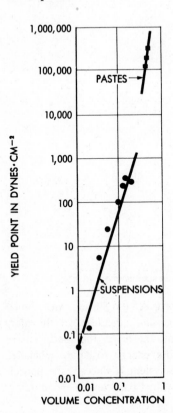

Fig. 7.13 Yield point of suspensions and pastes of no. 2 fraction of Fig. 7.11.

Water film in the plastic mass. If a plastic mass is made up with a monodisperse fraction of kaolinite when the size and shape of the particles are known, it is possible to calculate with considerable exactness the thickness of the water film between the particles. This is done by drying the mass and measuring the linear shrinkage. Taking a particular case, the particles are hexagonal plates, 0.6 micron across the flats and 0.05 micron thick. Statistically, these plates may be considered to be set up as in Fig. 7.14 to give 14 water films for each 1.88 microns of length. The linear shrinkage is 5.0 percent. Therefore, the film thickness is given by

$$\frac{\% \text{ linear shr. of mass} \times 1 \text{ cm}/100}{\text{No. of films per cm}} = \frac{0.050}{\frac{14}{1.88} \times 10^4} = 6.6 \times 10^{-7} \text{ cm}$$

$$= 6.6 \times 10^{-3} \text{ micron.}$$

Of course the film thickness will change with the pressure used in forming the mass, as will be discussed in the next chapter.

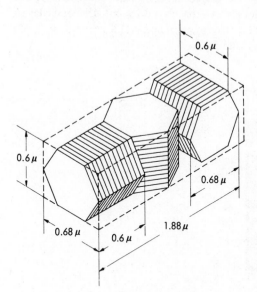

Fig. 7.14 Packing of kaolinite plates in a plastic mass.

Forces acting between the particles. In a clay mass at equilibrium the forces holding the clay particles together must be just balanced by those holding them apart. It is not difficult to measure the repulsion forces for any given separation distance. This is done by pressing the plastic clay between permeable pistons. The clay mass, after pressing, is dried and the water film thickness determined as described previously. With the size of each kaolinite particle known, the force between particles is readily çalculated, as will be done later in this chapter.

It is evident that the force holding the particles together is just equal to the repulsion force. For example, when a piece of plastic clay is pressed between permeable pistons at high pressure, water is squeezed from the mass and the particles come together to the point where their repulsion balances the external pressure. Now if the pressure is released under conditions where no water is available to flow back into the mass, then the films remain at their high-pressure thickness. There must be available for holding the particles together some force of the same magnitude as the released pressure.

The only force that would seem to be available is the capillary forces at the surface of the clay mass. It will be easier to understand this force if we consider the very simple case of two kaolinite plates surrounded by a water film. These plates, drawn to scale are shown in Fig. 7.15(a) for a thick water film and in Fig. 7.15(b) for a thin film. The radius of curvature of the capillary surface enables the attractive force to just balance the repulsion. The water in the films must be in hydrostatic tension. That this is actually so is indicated by the experiments of Westman, who showed values as high as 880 lb/sq in. for plastic ball clays.

The same reasoning applied to the two kaolinite plates may also be applied to a clay mass with millions of particles. The surface layer of water acts as a stretched membrane, forcing the particles together. As the clay dries out, the water layers between the particles decrease and the surface membrane becomes thinner and pulls down between the particles to exert greater force.

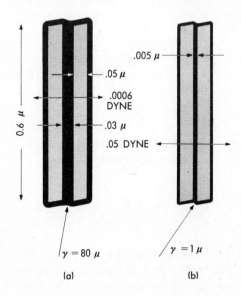

Fig. 7.15 Forces between a pair of kaolinite plates.

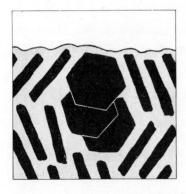

Fig. 7.16 A cross section of a clay paste at three different water contents.

Three consistencies of a clay-water mass are shown in Fig. 7.16, which is a cross section at the surface. A very simple experiment will illustrate the stretched membrane. A toy rubber balloon is filled with dry, pulverized clay, in which case the clay feels like a dry powder. If the balloon is slowly evacuated so that the atmosphere presses on the rubber to hold the clay particles together, a remarkable change occurs; the clay in the balloon now feels just like a plastic clay-water paste.

To sum up, a plastic mass has a yield point that must be exceeded before plastic flow may be initiated. When flow starts it may be carried to some extent before fracture occurs. The more plastic masses have both a high yield point and a long extension before fracture. The yield point is believed to be due to the holding together of the mass by the outer surface film and to the extensibility caused by platelike particles that readily slide over each other, perhaps lubricated by the water films.

6. Measurement of Flow

The flow properties of plastic masses may best be measured on the shear tester shown in Fig. 7.17. This machine shears a bar 1 cm square at any desired rate and plots the stress-strain diagram as shown in Fig. 7.6. A detailed drawing of this machine may be found in *Fine Ceramics*, p. 135.[1]

The flow properties of slips may best be measured by a revolving cup viscosimeter as shown on p. 116 of *Refractories.*[1]

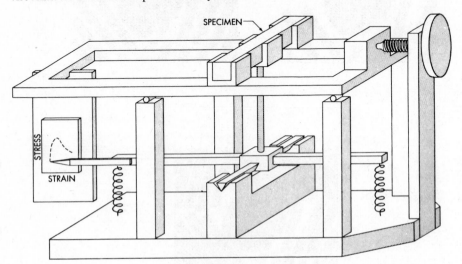

Fig. 7.17 Apparatus for measuring the flow properties of clay pastes in shear.

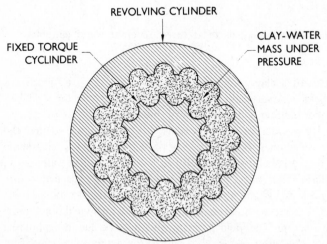

Fig. 7.18 Cross section of a device to measure shear in clay pastes of soft consistency.

---

[1]F. H. Norton, *Fine Ceramics* (McGraw-Hill, New York, 1970) and *Refractories* (McGraw-Hill, New York, 1968).

There is a kind of "no man's land" where the consistency of the mixture is between a slip and a plastic mass, a condition in which it cannot be handled satisfactorily. One way of testing in this region is to force the mix between a pair of concentric cylinders, the outer one of which revolves and the inner one measures torque (Fig. 7.18).

7. Systems with Nonclay Materials

Clay-grog-water system. Refractories and many clay products are made from mixtures of clay and grog. The plastic masses of this type have much the same flow properties as the clay-water system except that the grog increases the yield point, because the grog content is usually less than the clay and only acts as a filler.

Grogged bodies can be used as casting slips with water contents as low as 10%. These slips are usually thixotropic and must be vibrated to keep them from setting up.

Particle packing. As clay alone often has a large drying shrinkage and undesirable firing properties, it is common practice in many branches of the ceramic industry to add a nonplastic like silica, feldspar, or grog (hard fired clay). The particle size of the nonplastic has an important influence on the properties of the body, so some space will be devoted here to the influence of particle size distribution.

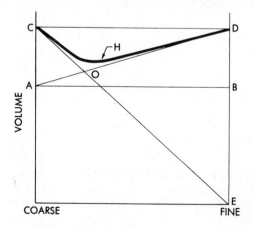

Fig. 7.19 Packing diagram of coarse and fine grog.

If the nonplastic is of one grain size, the volume of pores for crushed grains will be close to 45 percent of the bulk volume. In some cases it is desirable to have a lower pore volume in order to get the maximum possible amount of nonplastic in a given volume. By using two sizes of grog mixed together, the pore volume may be reduced as shown in Fig. 7.19. Line *AB* represents the true volume of 100 gm of nonplastic that consists of various mixtures of fine and coarse. *D* will be the bulk volume of the fine fraction and *C* of the coarse fraction, and line *CD* represents the bulk volume of the unmixed fine and coarse components. However, if the two are thoroughly mixed, the bulk volume will shrink to line *COD*, since the fine particles will fit into the pores of the larger ones. The minimum volume comes at *O*, where the pores in the larger fraction are just filled by the smaller one. The student should thoroughly understand this diagram and know, for example, the reason that line *CO* is prolonged to the base line while line *DO* is prolonged to the true volume line. It is helpful to take some 250-cc graduates, fill them with mixtures of two sizes of crushed grog, and then vibrate them until the bulk volume becomes a minimum. The packing of the particles can readily be seen during this operation.

The diagram of Fig. 7.19 is made up for two fractions with an infinite ratio between the fine and the coarse. Actually a ratio of ten is commonly attained, which gives a bulk volume curve shown by curve *H*.

If three components are used, still closer packing may be attained; the fine fraction goes into the pores of the medium fraction, which in turn fills the pores of the coarse fraction. Theoretically, still denser packing could be obtained by four or five components; actually, this does not follow, since it is impossible to keep a large size ratio between components if more than three are used.

Figure 7.20 shows the packing density for three sizes of crushed grog mixed in all proportions. The densest packing comes at 50 percent C, 10 percent M, and 40 percent F, with a porosity of 22 percent based on the bulk volume, not a great gain over the 25 percent figure for the fine and coarse mixture alone.

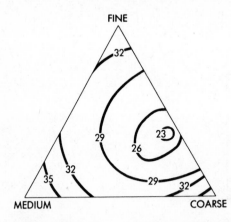

Fig. 7.20 Packing diagram of a three-component grog system. Contours represent porosity based on bulk volume.

Practical application of particle packing. So important is this question of particle sizing in the refractories industry that many modern plants separate the crushed grog into size fractions by screening and then combine these in definite proportions, taking great care that they do not unmix in handling or storage. Of course, the clay might be considered as a fourth component, for it fills the remaining pores.

The Scheidhauer and Giessing process for making large refractory shapes with no drying shrinkage and very little firing shrinkage consists of carefully preparing a closely packed grog and then coating the grog grains with a small amount of deflocculated slip. When the mass is pressed together by heavy ramming, the clay is forced into the few voids left. The grog particles make such close contact that little or no subsequent shrinkage can result.

The whiteware body is, perhaps by accident, a closely packed three-component system with the flint and feldspar as the coarse, the kaolin as the medium, and the ball clay as the fine component. It has been shown that molded densities even higher than those found in practice may be attained by a more careful selection of the components, thus reducing shrinkage and warping in the kiln.

References

Marks, L. S., *Standard Handbook for Mechanical Engineers*, 7th ed., McGraw-Hill Book Co., New York, 1968

Goodeve, C. F., A General Theory of Thixotrophy and Viscosity, *Trans. Faraday Soc.* **35**, 342, 1939

Scott-Blair, G. W., *A Survey of General and Applied Rheology*, 2nd ed., Pitman, London, 1949

Weymouth, J. J., and W. O. Williamson, The Effect of Extrusion and Some Other Processes on the Microstructure of Clay, *Am. J. Sci.* **251**, 89, 1953

Williamson, W. O., Effects of Rotational Rolling on the Fabric and Drying Shrinkage of Clay, *Am. J. Sci.* **252**, 129, 1955

Bloor, E. C., Plasticity: A Critical Survey, *Trans. Brit. Ceramic Soc.* **56**, 423, 1957

Astbury, N. F., A Plasticity Model, *Trans. Brit. Ceramic Soc.* **62**, 1, 1963

Williamson, W. O., Structure and Behavior of Extruded Clay. I, *Ceramic Age* **82**, (2) 39, 1966

Williamson, W. O., Structure and Behavior of Extruded Clay. III, *Ceramic Age* **82**, (4) 41, 1966

Ryan, W., The Deflocculation of a Blue Ball Clay, *Trans. Brit. Ceramic Soc.* **67**, 15, 1968; **69**, 33, 1970

Joyce, I. H., and W. E. Worrall, The Adsorbtion of Polyanions by Clays and Its Effect on Their Physical Properties, *Trans. Brit. Ceramic Soc.* **69**, 211, 1970

Goodwin, J. W., Rheological Studies on Dispersion of Kaolinite Suspensions, *Trans. Brit. Ceramic Soc.* **70**, 65, 1971

8

Forming methods

1. Introduction

It will be found that the ceramic industry employs many methods of forming ware. This is partly due to variations in the body, in part to the shape or size of the article being produced, and also to customs in the various branches of the industry. Table 8.1 shows how the various methods may be classified as to water content. In this chapter the usual forming methods will be discussed, in addition to a few others still in the experimental stages.

Table 8.1 Classification of forming methods

Forming process	Pressure used, psi	Water content, %
Slip casting	0^1	12–25
Plastic forming	5–50	25–30
Extrusion forming	50–10,000	15–20
Dry pressing	1000–15,000	5–10
Dust pressing	3000–20,000	$0–2^2$
Isostatic pressing	5000–100,000	0–15

[1]Higher for pressure slip casting.

[2]Plus binder and lubricant.

2. Body Preparation

The earliest method of preparing bodies for the forming operation was by treading under foot, but today there are many methods used to combine the raw materials into a workable body. In this section some of these processes will be described.

Refractories and heavy clay products. These bodies are combinations of clays and grog molded as soft mud, stiff mud, or dry pressed mixes. Soft mud mixes are prepared in a wet pan or pug mill, whereas the stiff mud mixes may be worked up in a pug mill or auger. Today, however, the large bulk of the refractories are made by the dry pressing method in which the body is prepared by a special mixer just before pressing. A few heavy refractories are made by slip casting, using a very heavy slip blunged at low speeds.

Fine ceramic bodies. Casting slips for these bodies are often made by the classic method of Fig. 8.1(a). Recent advances in filter presses allow for increasing the usual slip pressure of 125 psi up to 1000 psi, which gives a cake of lower water content and much faster dewatering.

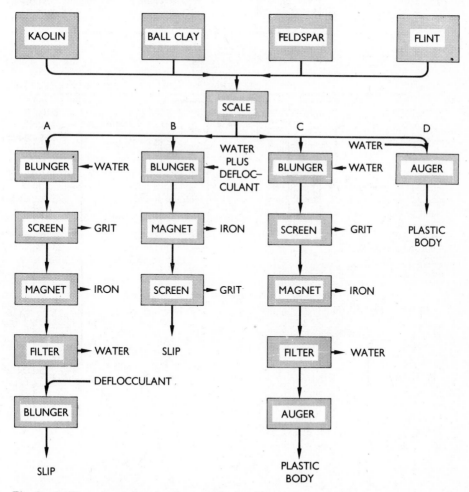

Fig. 8. 1 Flow sheet showing four methods of body preparation.

Some industries, particularly sanitary-ware manufacturers, make slip directly as in Fig. 8.1(b). This is possible because screens are now available to handle slips with specific gravity up to 1.85.

Plastic bodies can be made by the process of Fig. 8.1(c) using a vacuum auger. On the other hand with pure starting materials the body may be made directly as in Fig. 8.1(d).

The classic method of preparing dry press whiteware bodies is shown in Fig. 8.2(a). If washed or air-floated clays are used, direct mixing (Fig. 8.2(b) may be

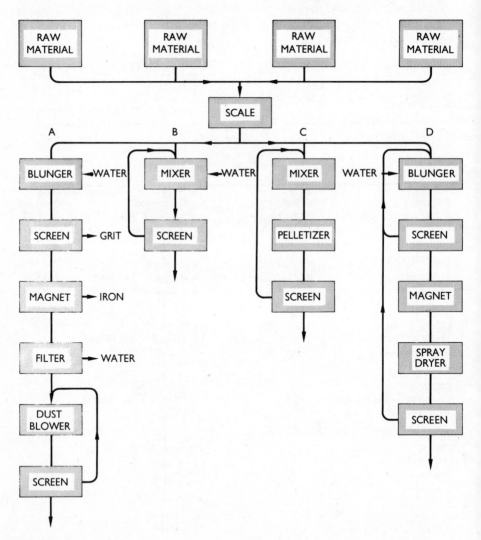

Fig. 8.2 Flow sheet showing four methods of making dry press bodies.

used. The method shown in Fig. 8.2(c) produces small spherical particles of batch in a pelletizer. Such particles have an advantage over irregular particles as they flow into the mold more evenly. A method in which a spray dryer produces small spheres is now used a great deal for tile, low-tension electrical porcelains, and fine-grained refractories; this is shown in Fig. 8.2(d).

3. Slip Casting

This process may be divided into two classes: (1) drain casting, in which the slip is poured into the mold, left a short time, and then drained out, leaving a thin shell against the inside of the mold; and (2) solid casting, in which the mold is filled with a slip and left until it casts into a solid piece.

The slip-casting method is much used in ceramic production as it is possible by this means to reproduce very complicated shapes in plaster molds. Obviously the slip will need somewhat different properties for solid casting than for drain casting, although there are some cases, particularly in sanitary ware, where both drain and solid casting occur in the same piece.

Slip making has been, and to a large extent still is, an art. We know what properties are desired but often are at a loss to know how to get them. It may be helpful to look at Table 8.2 to see what slip properties are wanted.

The casting slip is a suspension of clay, nonplastics, or both, in water. In order to obtain a stable, quick-casting slip it is necessary to have a high specific gravity together with a pourable condition and this requires a suitable deflocculant. The usual deflocculants are a combination of sodium silicate and sodium carbonate, but others such as tannates, humates, lignin, derivatives of sulfonic acid, ammonia, and polyphosphates are used.

Table 8.2 Desirable properties of casting slips

Properties	Drain casting	Solid casting
Low enough viscosity to drain well	x	
Good stability on storing	x	x
Ability to drain cleanly	x	
Produces sound solid casts		x
A low rate of settling on standing	x	x
Quick mold release	x	x
Fast casting rate	x	
Gives low shrinkage in cast	x	x
Gives high green strength of cast	x	
Gives high extensibility in cast	x	
Gives freedom from pin holes in cast	x	x
Gives freedom from scumming and wreathing	x	
Low degree of thixotropy	x	
Allows trimming without tearing	x	
High degree of thixotropy		x

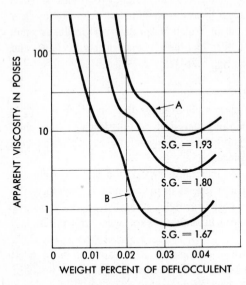

Fig. 8. 3 Viscosity curves for a whiteware slip. Solid casting would be made at *A* and drain castings at *B*.

The student may discover the correct amount of deflocculant by the following procedure. Make up a series of slips varying independently the amount of deflocculant and specific gravity. These slips should be blunged with a high-speed mixer and aged for a few hours. Then they may be tested for apparent viscosity to produce a set of curves like those in Fig. 8.3. Solid casts should be made from slips around point *A* and drain casting should be around point *B*. The slips in the region of *A* and *B* may then be cast in the test molds to ascertain the exact condition that will give the optimum properties of casting rate and firmness of casting.

Whiteware slips. These slips are used with a specific gravity of 1.75 to 1.95 for an average body composition of 52% nonplastic, 30% kaolin, and 18% ball clay. The viscosity of drain-cast slip should be between 1 and 5 poises and that for solid castings should be 5 to 50 poises. The deflocculant used is generally S-brand silicate of soda and sodium carbonate added in amounts that together comprise about 0.05% of the slip weight. The action of the deflocculant is greatly influenced by the soluble salts brought in by the ball clays and water. In general, low-soluble salts are desirable to give a firm cast. Some of the large potteries use an elaborate computer control to keep additions of deflocculants and other materials at the proper point.

The steps in drain casting are illustrated in Fig. 8.4. The casting time may vary from 10 minutes for thin casts up to several hours for a half-inch wall. A proper slip drains cleanly, leaving a perfectly smooth inner wall.

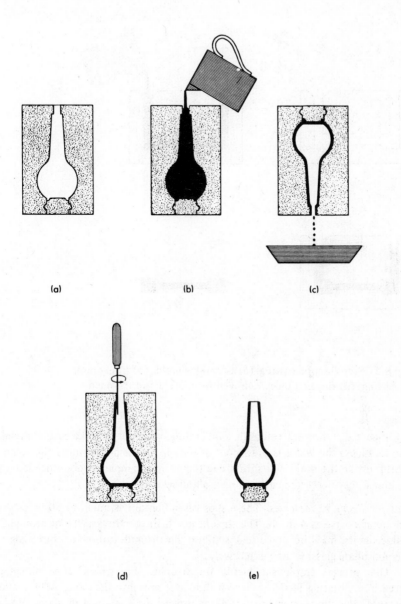

Fig. 8.4 Steps in drain casting: (a) assembled mold;
(b) pouring slip and casting; (c) draining; (d) trimming;
(e) removing mold.

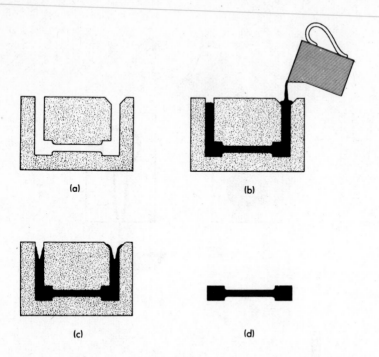

Fig. 8.5 Steps in solid casting: (a) assembled mold; (b) filling mold;
(c) casting; (d) finished piece removed from mold and trimmed.

Figure 8.5 shows the steps in solid casting, which is very like casting metal in a sand mold. As the water is drawn out of the slip by the plaster mold the volume of the slip inside the mold will shrink and therefore an ample supply is needed during the time of casting or the cast will have a hollow core.

Slips for heavy refractories. These slips often contain as much as 50 or 60% grog and are as coarse as 4 mesh. The slips have a high specific gravity of around 2.00. Deflocculants may be the usual sodium silicate and carbonate, but Calgon (a polyphosphate glass) is used extensively.

One process employs a highly thixotrophic slip, which, after blunging, is placed in a vibrating feeder to keep it fluid as it runs into the mold. After resting in the mold, the slip sets up into a solid with very little removal of water. For glass refractories the slip is sometimes evacuated during blunging to remove all the entrapped air.

Pressure casting. It has been the practice for some time to make solid castings with the slip under an atmosphere of pressure to double the casting rate. Recently there have been attempts to still further speed up the rate by using higher slip pressure — up to 1200 psi. At a slip pressure of 1000 psi and a temperature of 80°C a

½-in.-thick wall can be cast in one minute and the resultant cast is found to have a much lower drying shrinkage than normal. A normal time to cast the ½-in. wall would be one to two hours, so a tremendous saving in casting time has been achieved. Of course plaster molds cannot be used under the high pressure and temperature, so new materials must be found such as porous plastics. This is really a filtering process and will be a great help in setting up a completely mechanized casting unit.

The mechanism of casting. The process of casting in plaster molds is still not completely understood. What seems to happen is that the fine pores in the plaster draw the water from the slip by capillary forces, thus lowering the water content of the slip adjacent to the plaster until a rigid layer of the body is built up which slowly thickens as the water moves through the cast layer. The rate of building up the rigid layer falls off with time because the plaster is absorbing water and its capillary pull is therefore decreasing, and at the same time the thickening layer becomes less permeable. It is found that the rate of layer build-up falls off as the square of the time. For example, if a layer is 0.1 in. thick in 3 minutes it will reach 0.2 in. in 9 minutes. The distribution of water during a casting process at specific times is shown in Fig. 8.6.

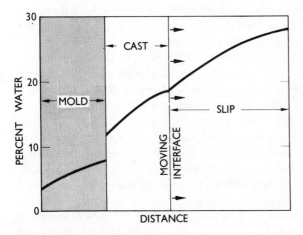

Fig. 8.6 Water distribution during the casting operation.

Plaster molds. There is not space here to describe making the molds, a process which must be learned from actual contact with an expert mold maker. However, a few modern developments may be mentioned. The plaster mixing is now closely controlled as to time of soak, time of mixing, and temperature. The mold shop temperature and humidity is also controlled in some cases. Some potteries feel it worthwhile to de-air the plaster during mixing to eliminate pin holes and give a longer mold life. The case molds which used to be of plaster are now made of a much lighter, more durable reinforced plastic. Plaster casting molds last for 50 to 300 pourings.

4. Forming from Soft Plastic Masses

This is the earliest method for forming ceramic articles such as pots and brick, and even now some pottery is thrown on the wheel and brick is formed in wooden molds. In present-day production the soft plastic method is used for soft mud bricks, jiggered or roller-formed tableware, hot pressing insulators or cup handles, and forming by the ram process.

Soft mud brick. Although such brick are still being hand molded and sun dried in some smaller brick yards, the bulk of the production is carried out with a machine to fill the molds. The clay and water is introduced into a sort of pug mill which continuously forces it into a die chamber, where a plunger forces it down through a multiple die into a wooden five-part brick mold indexed below it. The mold moves along, is turned over, and the brick are dumped on a pallet for transfer to the dryer. The mold is washed, dusted with dry sand, and returns to complete the cycle. About 30,000 brick can be turned out by such a machine in an eight-hour day. These brick are called "sand struck."

Jiggering. This process is used largely in the whiteware industry to form plates and some types of hollow ware. Until recently hand jiggering was the rule, but now most potteries use automatic jiggers for everything except a few specialties.

The operation starts out with a lump of plastic body of the proper weight. This is formed into a round bat like a pancake by striking with a plaster tool or by spreading it on a revolving disk with a descending tool. As this is the only part of the process where the body is appreciably deformed, it is vital that the finished bat be completely homogeneous. The bat is next transferred to a plaster mold shaped like the upper surface of a plate, for the plate is formed upside down, as shown in Fig. 8.7. This mold is then set in a chuck on the upper end of a vertical shaft that is turning at the rate of 300 to 400 rpm. A tool contoured like the underside of the plate is pulled down to make contact with the bat, now lubricated with water, and forms the surface precisely, partly by scraping off excess body and partly by forcing the body down on the mold. A section of the tool in Fig. 8.8 shows this operation. The plaster mold and plate are finally taken from the jigger chuck and sent through a continuous dryer.

The jiggering operation requires great skill to form large plates free from warping, for any unequal strain at any stage of the process will show up in the firing. The orientation of the clay particles on the jiggered surface also contributes to the smooth finish attained in this operation.

It was long thought that such an exacting operation could not be done by machine, but as early as 1935 an automatic jigger for cups was in use in Sweden. About the same time an automatic jigger of high capacity was developed by the

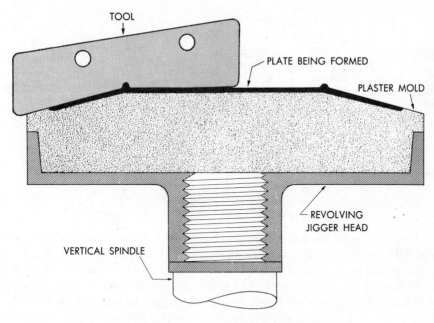

Fig. 8.7 Cross section of a jigger head as plate is being formed.

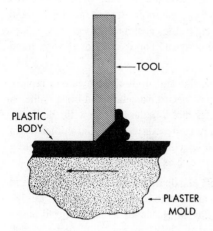

Fig. 8. 8 Operation of the jigger tool.

Homer-Laughlin Company of Newell, West Virginia, for flat ware. During this period W. J. Miller developed and put on the market an automatic jigger that is now used in many potteries.

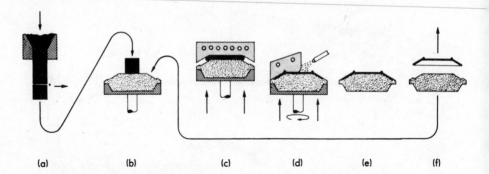

(a)   (b)   (c)   (d)   (e)   (f)

Fig. 8. 9 Operational sequence of the Miller jigger.

The Miller jigger operates as shown in Fig. 8.9. A round column of de-aired plastic body is fed to the machine, where it is cut off in correct lengths and dropped on the jigger mold. This mold is then pressed up against a heated die that spreads the body on the plaster mold to almost its final size. The die is heated so that the resulting film of steam will prevent sticking of the body. The mold and formed body are then placed on the jigger head where they are held by vacuum. The jiggering operation takes off little clay, but produces a smooth finish because the surface is sprayed with a water mist. The jiggering speed is much higher than in hand operation, 500 to 1200 rpm. The mold and finished plate are then carried to a continuous dryer, where the plate separates from the mold. It is then removed and the edges are trimmed. The capacity of an eight-line machine is about 600 dozen per hour; consequently it saves a great amount of labour compared with hand operation.

Roller forming. This method of forming both flat and hollow tableware is rapidly replacing the jigger even for tender bodies such as hard porcelain and bone china. A bat of soft plastic clay is applied to the plaster mold just as is done with the jigger, but instead of a blade, a polished (and often heated), contoured roller comes down and rolls out the body (Fig. 8.10). These machines can be made completely automatic and have about the same production rate as the jiggers.

Hot pressing. This term applies to pressing soft plastic mixes into a heated mold and should not be confused with the same term applied to pressing refractories at high temperatures. Here a lump of soft plastic clay is placed in a plaster mold and a heated, revolving tool is pressed down to form an object such as an insulator. Small pieces, such as cup handles, are impact-pressed between two heated steel molds.

The Ram process. In this operation a lump of soft plastic body is pressed between two plaster molds to form a plate or other object. The water removed by the molds may be periodically blown out to prepare for the next cycle. The operation is shown in Fig. 8.11. The advantage of this process is that pieces more complicated than those with surfaces of revolution can readily be formed.

Fig. 8.10   Automatic roller machine for forming plates. (Homer Laughlin China Co.)

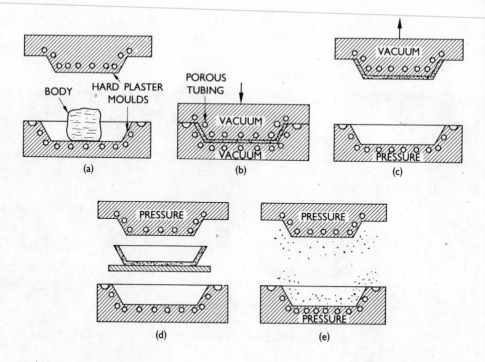

Fig. 8.11 Operation of the Ram method of forming ware: (a) slug of plastic body in mold; (b) pressing the piece; (c) lifting the piece up with the top mold; (d) releasing the piece by air pressure; (e) blowing moisture from the mold.

5. Stiff Plastic Forming

Bodies with this consistency are formed by extruding through a die, continuously by an auger, or intermittently by a plunger. Long pieces of simple or complicated cross section may be made in this way and later cut into short sections for brick, block, or pipe.

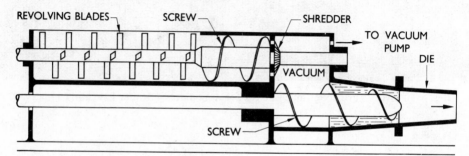

Fig. 8.12 Cross section of a vacuum auger.

Auger extrusion. Figure 8.12 shows a cross section of a vacuum auger with the parts labeled. The mix is fed into the top where it is kneaded by the blades of the central shaft and at the same time forced through a series of small openings into the vacuum chamber. There it is picked up by another auger shaft and forced through the die, which is often heated and lubricated to reduce friction. The column from the die is carried on a moving belt or carriage and cut into the desired length, often by one or more stretched wires.

Inserts may be placed inside the die to produce hollow columns such as drain tile or perforated brick. The auger can be a high-production machine turning out as much as 10 tons per hour.

Slugs for high-tension electrical insulators are made on augers that turn out pieces as large as 3 ft in diameter.

Piston extrusion. In this operation a cylinder is filled with shredded, stiff plastic body, which is forced out through a die to form a column of uniform cross section that may then be cut into the desired lengths. Sometimes the cylinder is evacuated after filling. This process gives a more uniform column than the auger, but there is still a density difference between the exterior and center of the column.

Piston extrusion is used for fine-grained refractories and electronic bodies, in which case plastizers, such as methocellulose and polyvinyl alcohol, are used. The body may be placed in the cylinder as a preformed slug and forced through the die at pressures up to 5000 psi. Another method adds the shredded body with a wax binder to the cylinder, evacuates the cylinder, and forces the body through the die. Pieces as small as 1 mm diameter with six 0.1-mm holes may be made in this way.

Larger sewer pipes are made in a vertical piston extruder. The evacuated cylinder is filled with clay from an auger, a steam-driven piston forces the clay through an annular die into a fixture to form the bell, after which the length of pipe is extruded and cut off with a wire.

6. Dry Pressing

This method of forming is used by the refractories manufacturers to produce at least 85% of the fireclay brick and nearly all silica and basic brick. All wall tile, floor tile, and some quarry tile are dry pressed and most low-tension electrical porcelain is formed in this way.

Pressing refractories. The batch of plastic fireclay and grog or flint clay is mixed with 7 to 10% water to the consistency of damp powder. This batch is fed into an automatic toggle press with a four-brick die box, giving pressure up to 15,000 psi at a rate of 2000 brick per hour. Hydraulic presses and hydraulic toggle presses are also used to a limited extent. A toggle press making refractory shapes is shown in Fig. 8.13.

The same types of press may be used for less plastic mixes such as ganister, magnesite, and chromite, usually with the addition of an organic binder such as dextrine.

Fig. 8.13 Toggle press forming refractory blocks. (Chisholm, Boyd and White Co.)

The maximum useful pressure possible in the toggle press is limited by laminations that are formed in the bricks by expansion of trapped air when the pressure is released. Attempts to help this situation are:(1) to evacuate the mold box at each stroke, and(2) to use a double stroke, the first at medium pressure to force out the air and the second at high pressure to compact the structure.

Fig. 8.14 Automatic tile press. (John Weiland and Son Machine Shop)

Since the wear on the die box is severe when pressing high-grog mixes, hard metals such as alloy steels or stellite are used. The mold liners can be reground a number of times with appropriate shims placed behind them. When extremely abrasive mixes are pressed, tunsten carbide die boxes are used, which give 75,000 to 100,000 cycles before replacement.

Forming tile by dry pressing. Today most tile bodies are prepared as described at the start of this chapter and fed into automatic presses of various types. The screw press seems to be preferred as the impact of the upper platen causes the mix in the die box to flow into an even structure. Usually the double stroke described in the previous section is used. An automatic press (Fig. 8.14) will press six 4¼ × 4¼ in. tiles per stroke at a rate of 30 strokes a minute.

Dry pressing low-tension electrical insulators. This product is made from a triaxial body, with the batch prepared as described in the first section of this chapter. Screw and toggle presses have now been largely displaced by completely automated hydraulic presses such as shown in Fig. 8.15. The molds are made of case-hardened steel or heat-treated, carbon-chrome air-hardening steel. Mold life goes up to 500,000 pieces as the pressures are relatively low, 500 to 1500 psi.

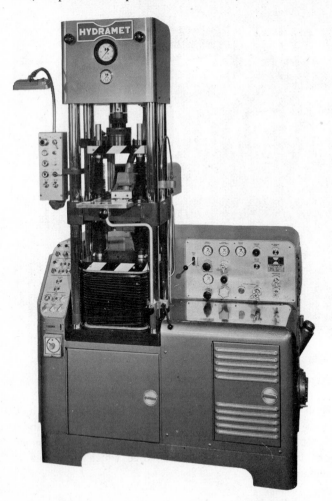

Fig. 8.15 Automatic hydraulic press for rapid production of small pieces. (Hydramet American, Inc.)

7. Dust Pressing Electronic Ceramics

This process, employed largely for the steatite porcelains, uses only 1 to 2% water in the mix and higher pressing pressures 5000 to 10,000 psi. Organic binders such as a hard wax emulsion, dextrine, or polyvinyl alcohol are needed to give good flow in the mold. The molds and presses are similar to those described in the previous section.

8. Isostatic Pressing

This method of forming consists of enclosing a preform or a dry powder in a rubber mold, then subjecting the outside of the mold to hydrostatic pressure from 5000 to 20,000 psi (Fig. 8.16). The advantages of the process are 1) no water or binder is needed in the mix, 2) a high green density is produced which reduces the firing shrinkage, 3) the green density is substantially uniform throughout the piece causing a minimum of warping in the kiln, and 4) the high forming pressure aids sintering, so that firing temperatures may be reduced. On the other hand, the surface against the rubber mold cannot be held to close tolerances.

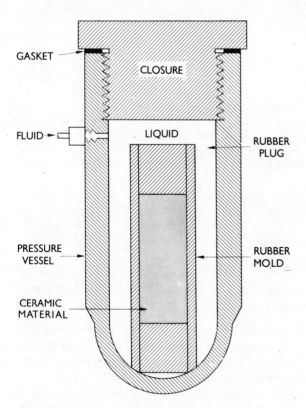

Fig. 8.16 Wet-bag isostatic pressing equipment.

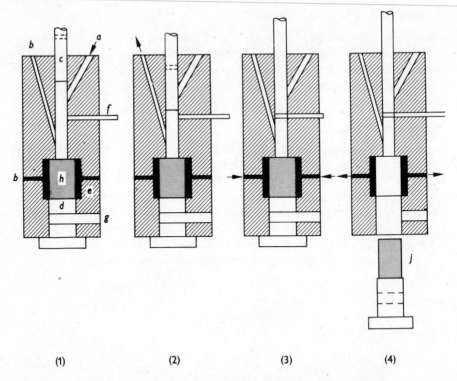

(1)   (2)   (3)   (4)

Fig. 8.17 Automatic dry-bag isostatic press: (1) filling cycle with vibration; (2) evacuation; (3) fluid pressure; (4) ejection. The parts are identified as follows: a, feed tube; b, vacuum tube; c, upper closure; d, lower closure; e, rubber mold; f, upper closure lock; g, lower closure lock; h, mold fill; i, inlet for pressure fluid; j, finished cylinder.

The most important use of this process is for forming spark-plug cores of alumina. By using the rubber as part of the isostatic medium a rapid automatic process is possible (Fig. 8.17). Other uses for this process are the fabrication of sewer pipes and zircon paving blocks.

9. Green Finishing

The finishing is an important step in making high-grade ware. A few of the more common processes will be outlined here.

Trimming. Jiggered ware must have the feather edge on the rim trimmed off with a scraper and then sponged to make it smooth. Cast ware must have the top trimmed off, that is, the extra part that holds the slip for shrinkage, commonly called the spare. This trimming is usually done in the mold.

In many pieces there are seams left by the joints in the mold. In the case of earthenware these seams may readily be sponged off, but not so in vitreous ware.

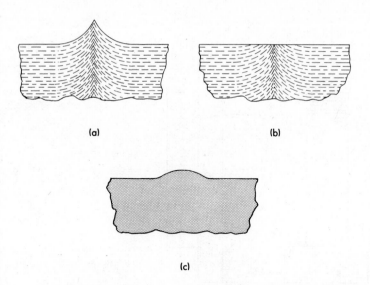

Fig. 8.18 Seam left when casting porcelain: (a) cross section of
cast piece at seam showing direction of kaolinite plates; (b) same
section scraped smooth; (c) same section after firing.

Here, no matter how perfectly they are leveled off in the dry state, they appear
again after firing. This is an interesting phenomenon, which may be explained by
the preferred orientation of the clay particles during casting somewhat as shown in
Fig. 8.18. Here it is seen that the particles at the seam are normal to the general
surface, while all the others are parallel to it. Therefore, the thickness shrinkage in
firing will be greater for the average area than at the seam, which thus is left
protruding. This is an excellent example of orientation influencing shrinkage. There
are a number of ways of getting around this trouble. One method used with bone
china consists of hammering the freshly cast piece with a small hammer at and
about the seam to cause plastic flow and give a random orientation. Another
method used with high-fire porcelain is to fire just below the maturing temperature,
grind the seam down with an abrasive wheel, and then refire.

Turning. Some fine ware, after forming and while partially dried (in what is often
called the leather hard stage), is placed in a chuck on a lathe or on a potter's wheel
and turned with steel tools to exact dimensions. The feet of cups are often turned
in this way. The eggshell porcelain of the Chinese was made by turning from heavier
thrown pieces, and this method was used by the ancient Greeks to form their
exquisite vases. Large high-tension insulators are now often turned out of a dried
blank on a lathe, as shown in Fig. 8.19. As nearly all drying shrinkage has taken
place before turning, very precise shapes may be achieved in this way.

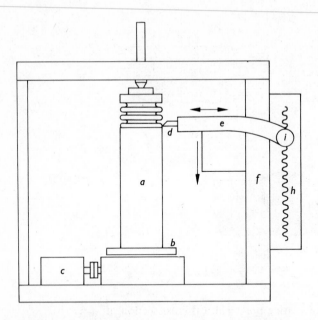

Fig. 8.19 Automatic lathe for turning insulator blanks:
a, blank; *b*, motor-driven revolving head; *c*, motor; *d*, cutting
tool; *e*, horizontal traversing arm; *f*, vertical slide; *h*, lined
template; *i*, photocell to follow line.

Burnishing. The glossy appearance of the ware made by our southern Indians is due
not to a glaze but to a burnishing operation, in which a polished pebble is rubbed
over the surface of the leather hard piece. The same method may be carried out
more expeditiously in the lathe, for example, in the case of Wedgwood jasper ware.
Here again the mechanism of burnishing is the production of a surface layer of clay
plates all laid down parallel to the surface like shingles on a roof. Even the firing
operation does not entirely destroy the polish.

Jointing or sticking up. Much ceramic ware is fabricated in pieces and then joined
together by using slip as glue. Examples are the handles on cups or a complex piece
of sanitary ware. Some porcelain figure groups are assembled from as many as sixty
pieces. The edges to be joined are roughened and then coated with slip and quickly
assembled. The important thing is that the two pieces to be joined have exactly the
same water content. If this condition is observed, there will be no differential
shrinkage in drying.

References

Hauth, W. E., Jr., Slip Casting of Aluminum Oxide, *J. Am. Ceramic Soc.* **32**, 394,
1949

Blackburn, A. R., Plastic Pressing, *Bull. Am. Ceramic Soc.* **29**, 230, 1950

Adcock, D. S., and I. C. McDowall, Mechanism of Filter Pressing and Slip Casting, *J. Am. Ceramic Soc.* **40**,335, 1957

Thurnauer, H., *Controls Required and Problems Encountered in Production Dry Pressing,* Wiley, New York, 1957

Herrmann, E. R., and I. B. Cutler, The Kinetics of Slip Casting, *Trans. Brit. Ceramic Soc.* **61**, 207, 1962

vanWunnik, J., Practical Control of Slip Properties, *Ceramic Age* **78** (12), 45, 1962
Phelps, G. W., and J. vanWunnik, Theory and Practice of Slip Control, *Ceramic Age* **78** (17), 35, 1962

Wehrenberg, T. M., *et al.*, Isostatic Pressing Large Refractory Blocks, *Bull. Am. Ceramic Soc.* **47**, 642, 1968

Ovenston, A., and J. J. Benbow, Effect of Die Geometry on the Extrusion of Clay-like Material, *Trans. Brit. Ceramic Soc.* **67**, 543, 1968

Everett, D. W., Isostatic Pressing – Revolution in Refractory Forming, *Ceramic Age* **85** (8),14, 1968

Dal, P. H., and W. J. H. Berden, The Capillary Action of Plaster Molds, in *Science of Ceramics*, Vol. 4, p. 113, British Ceramic Society, 1968

Gill, R. M., and J. Byrne, Application of Isostatic Pressing Techniques to the Production of Dense Ceramic Bodies, in *Science of Ceramics*, Vol. 4, p. 91, British Ceramic Society, 1968

Ayer, J. E., and F. E. Soppet, Vibratory Compaction. Part I. *J. Am. Ceramic Soc.* **48**, 180, 1965; Part II. *J. Am. Ceramic Soc.* **49**, 207, 1966; Part III. *J. Am. Ceramic Soc.* **52**, 414, 1969

Reingen, W., Durability of Moulds, *Ceramics* **21**, 6, 1970

9

Drying ceramic ware

1. Introduction

The drying process is an important step in the manufacture of many ceramic articles. Although the dictates of economy require the fastest possible drying, too fast a schedule causes differential shrinkage of such magnitude as to produce cracking. In this chapter the principles of the drying of porous solids will be discussed.

It is assumed that the student is already familiar with the properties of air and has some knowledge of psychrometry; if not, this knowledge may be obtained from the references at the end of the chapter. However, it should be kept in mind that moving air serves a twofold purpose in the drying process; it supplies heat to the ware as compensation for the evaporative cooling, and it carries away the water vapor formed.

2. Internal Flow of Moisture

Water evaporated from a piece of ware by drying must come mainly from the interior of the piece through the fine interconnecting channels.

Internal flow. The rate at which this water flows through a given structure, as shown in Fig. 9.1, is given by

$$\text{Volume rate of flow} = k\,\frac{\text{driving force}}{\text{flow resistance}},$$

or

$$\frac{dV}{dt} = \frac{k(C_2^1 - C_1^1)}{l} \cdot \frac{p}{\eta},$$

where dV/dt is the volume rate of flow, C_1^1 is the water concentration on the wetter face, C_2^1 is the water concentration on the drier face, k is a constant, l is the length of path, p is the permeability of the body, and η is the viscosity of water.

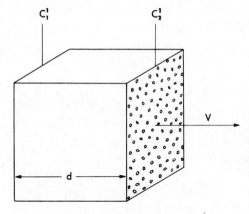

Fig. 9.1 Movement of moisture through a porous medium with a moisture gradient of $(C_1^1 - C_2^1)/d$.

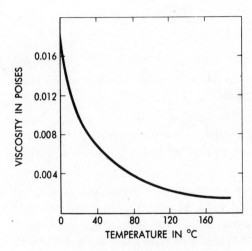

Fig. 9.2 Viscosity of water at various temperatures.

From the above relation it is evident that to increase the rate of water flow in a given material we have the choice of increasing the permeability, increasing the moisture gradient, or decreasing the viscosity of the water. Increasing the moisture gradient, as will be shown later, cannot be carried beyond a certain point without causing rupture of the body; on the other hand, the viscosity of the water may be decreased by working at higher temperatures (see Fig. 9.2), and the permeability increased by a coarser structure.

Moisture distribution. Many studies have been made of the moisture distribution in a drying solid. Figure 9.3 gives a typical distribution for a drying slab with no loss from the edges. The lines of equal moisture content at first are nearly flat but soon become highly curved. The moisture gradient is, of course, given by the slope of these lines. The more slowly the drying takes place, the slighter is the curvature.

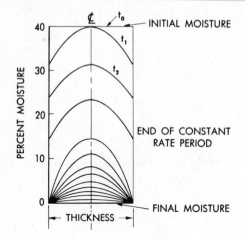

Fig. 9.3 Moisture distribution in a slab of drying clay.

3. Surface Evaporation

The rate of evaporation from the surface of drying clay ware depends on many factors which will be covered in this section.

Evaporation from a free water surface. The evaporation rate from a free surface is dependent on the air temperature, the air velocity, the water content of the air, and the water temperature. These factors are well known and can be found in treatises on drying.

Drying rates of ceramic bodies. Should the weight of a drying ceramic piece be plotted against time, the result would be a smooth curve without any great significance. However, if the rate of drying, or the slope of the weight-loss curve, is plotted against the water content of the piece, a curve like that in Fig. 9.4 results.

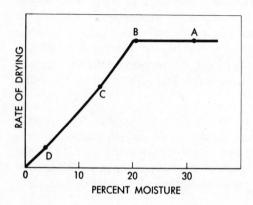

Fig. 9.4 Rate of water loss in drying moist clay.

As first pointed out by Sherwood, the wet clay starting at point *A* dries at a constant rate until *B* is reached. This constant rate of drying is about half that for a free water surface. At point *B* the rate starts to decrease rapidly and finally reaches the origin. It is significant that point *B* is that at which the mass changes from a dark to a light color. In other words, down to point *B* there is over the surface a continuous water film which acts as free water, but below this point the water retreats further and further into the pores so that the drying rate becomes less and less. These steps in the clay structure are shown by Fig. 9.5. In the next section it will be shown that this behavior is closely connected with the drying shrinkage.

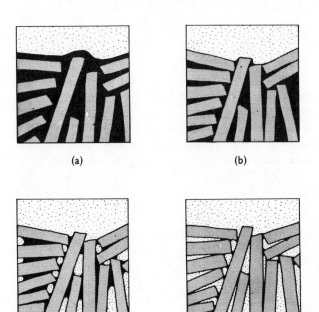

(a)   (b)

(c)   (d)

Fig. 9.5   Stages in drying moist clay. A cross section at the surface. The letters correspond to those on the curve of Fig. 9.4.

4. Drying Shrinkage

Mechanism of drying shrinkage. A plastic ceramic body may be dried slowly under conditions that permit a continuous measurement of the weight and the volume. From these data it is then possible to construct a curve as shown in Fig. 9.6 which tells an interesting story. The drying starts at *A* with a uniform volume decrease just equal to the water lost until point *B* is reached, after which no further volume change occurs. If the line *AB* is extended, it will pass through the origin at 45°.

The significance of this curve is not hard to see. Between *A* and *B* the water lost comes from the layers between the particles, so that the latter come closer and closer together until they touch at *B* and can consolidate no further. From *B* to *C* the bulk volume does not change and the water removed comes from the pores.

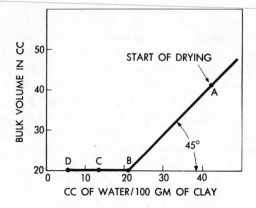

Fig. 9.6 Drying shrinkage curve of a moist clay. The letters correspond to those in Figs. 9.4 and 9.5.

Shrinkage curves for clays. Some clays depart from this curve to a slight extent (Fig. 9.7). Curve 1 is for a residual kaolin which shows a slight expansion at the very end of the drying period. There is no known explanation for this behavior. Curve 2 is for a ball clay which shows a slight secondary shrinkage as it reaches dryness. This is probably caused by a small amount of montmorillonite in the clay, which retains water between the lattice planes until the very end. The total shrinkage of a clay varies a great deal, the finer grain causing greater shrinkage.

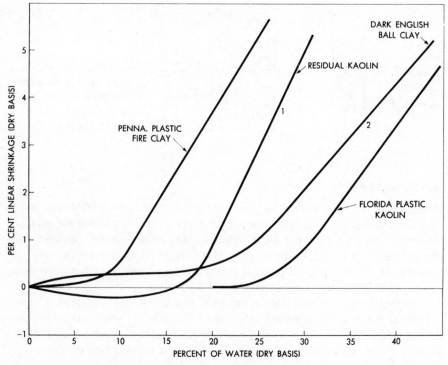

Fig. 9.7 Linear drying shrinkage curves for several clays.

Shrinkage conversion. It is often desirable to convert linear shrinkage to volume shrinkage or *vice versa*. When changes are very small the ratio is 1 to 3, but as they get larger, there is a considerable departure from this simple ratio. This is made clear in Fig. 9.8, where seven pieces are added to a unit cube to form a larger cube.

Let

b = volume expansion (% initial volume),
a = linear expansion (% initial cube edge),
1 = initial volume of the cube.

Then the final volume of the cube is

$$\frac{b}{100} + 1,$$

or, in terms of a, based on the sum of the volume of the seven pieces,

$$1 + 3\left(\frac{a}{100}\right) + 3\left(\frac{a}{100}\right)^2 + \left(\frac{a}{100}\right)^3 = 1 + \left(\frac{a}{100}\right)^3,$$

as

$$\frac{b}{100} + 1 = \left(1 + \frac{a}{100}\right)^3;$$

then

$$\sqrt[3]{\frac{b}{100} + 1} = 1 + \frac{a}{100},$$

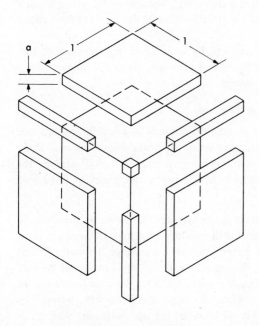

Fig. 9.8 The expansion of a cube.

$$\frac{a}{100} = \sqrt[3]{\frac{b}{100} + 1} - 1,$$

or

$$a = 100\left[\left(\sqrt[3]{\frac{b}{100} + 1}\right) - 1\right].$$

In the same way it can be shown that to change linear shrinkage to volume shrinkage based on initial values the following relation applies:

$$b = 100\left[\left(\frac{a}{100} - 1\right)^3 + 1\right].$$

It will be instructive for the student to derive the two other, similar relations based on final rather than initial values.

There is much confusion about the base of the shrinkage calculation. For example, if a piece of plastic clay 10 mm long shrinks 1 mm, the linear shrinkage will be 10.0 percent based on the initial length, but 11.0 percent based on the final length. The base should be stated in all cases. In this book the base is initial length unless otherwise stated.

Methods of reducing drying shrinkage. Excessive shrinkage is undesirable as it tends to cause cracking and distortion of the ware. The commonest cure is to add nonplastics to the clay. These relatively coarse materials simply reduce the number of water films per unit distance by displacing a group of clay particles and their associated films by an equal volume of stable particles. In the same way coarse-grained clays shrink less than fine-grained ones. Because of the orientation of the clay plates, cast bodies shrink less (in the direction of the surface) than plastic ones.

Drying shrinkage can be greatly reduced by molding at high pressure so that the water films are reduced to a low order. For example, many dry pressed bodies have a negligible amount of drying shrinkage.

5. Achievement of Maximum Drying Rate

In any manufacturing operation, the greater the amount of saleable product that can be turned out with a given piece of equipment, the lower will be the cost of the operation. This is true in drying, so every effort is made to dry ware at the maximum safe rate. Glass pots that formerly took months to dry may be dried with modern equipment in a few days.

Humidity drying. It was explained earlier in this chapter that lowering the viscosity of water by high temperatures speeded up the drying operation. However, in most cases, putting green ware into a high temperature would remove the water so fast that cracking would result. Therefore, the expedient of heating the ware all the way through in a saturated atmosphere was hit upon, for little water is lost from the piece under these conditions. Then, when the ware is thoroughly heated, the humidity is reduced as fast as possible without setting up dangerous stresses. A typical schedule for a humidity-controlled firebrick dryer is shown in Fig. 9.9.

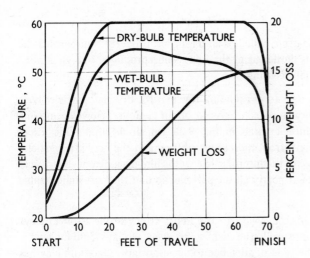

Fig. 9.9 Typical conditions in a humidity dryer.

Size factors. In the case of several objects of different sizes made from the same body, the larger ones will not only dry more slowly under a standard condition, but will also have a greater tendency to crack. Macey has shown that with cubes the safe drying rate is proportional to the edge of the cube.

Fig. 9.10 Kaolinite crystals after drying from an aqueous suspension to show preferred orientation. (Taken by C. E. Hall of Massachusetts Institute of Technology.)

6. Green Strength

Dry or green strength, as it is often called, is an important property that permits the handling of ware before it is hardened in the kiln. Some products, like glass pots, must stand rough handling before firing.

Mechanism of dry strength. The forces holding together the dry clay particles have been more or less of a mystery to those who write about ceramics. However, no one closely examining the kaolinite crystals in Fig. 9.10 can doubt that ionic forces hold them together. In this electron microscope photograph the kaolinite particles were dried down from a slip. As the inter-particle water film disappeared, the forces present were sufficient to align nearly all the particles so that their crystallographic axes were parallel.

Factors affecting dry strength. Clays vary greatly in dry strength; the stronger clays probably contain some montmorillonite, while the weaker clays contain coarse particles. It is known that the type of adsorbed ion is important. A sodium clay has about three times the strength of the equivalent hydrogen clay.

Many investigators have found a marked increase in dry strength of clays or clay-containing bodies when they are completely dried. For example, Pask shows that a kaolin in equilibrium at 100%, 50%, and 0% humidity gave strengths of 150 psi, 180 psi, and 260 psi, respectively.

7. Types of Dryer

Tunnel dryers. The heavy clay industry and many refractories manufacturers use a tunnel-type dryer, in which loaded cars pass slowly through the length of the tunnel (Fig. 9.11). The heat may be supplied by steam coils beneath the cars, by hot air from an air heater, or by waste heat from the kilns. These dryers are simple to construct but do not dry evenly from top to bottom of the car.

Cross-circulation dryers. These dryers circulate heated air laterally through the cars loaded with ware by means of fans or jets (Fig. 9.12). The cars may remain stationary during the drying period or move continuously through. With proper adjustment the drying is uniform over the whole charge.

Humidity dryers. These are usually tunnel dryers, but are divided into sections with independent heat and humidity controls (Fig. 9.13). In this way the ware can be heated up rapidly in a saturated atmosphere to reduce the viscosity of the water, then, while the temperature is maintained, the humidity is reduced as the ware passes from one section to another. There are also means for circulating air across the dryer. These dryers permit fast drying without cracking the ware.

Hot floors. Refractory shapes, sewer pipes, and other heavy pieces are often dried on a concrete floor with steam pipes or heating ducts embedded in it. Some hot floors have overhead heaters to blow hot air onto the ware to accelerate the drying.

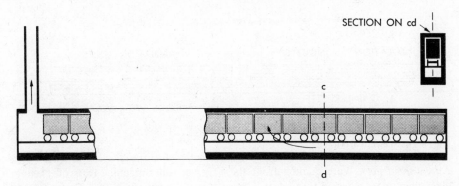

Fig. 9.11 A tunnel dryer.

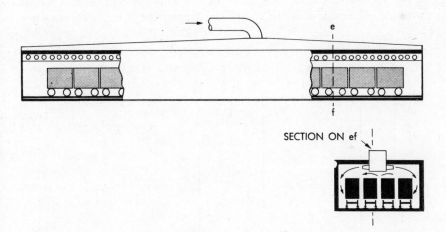

Fig. 9.12 A dryer with cross circulation.

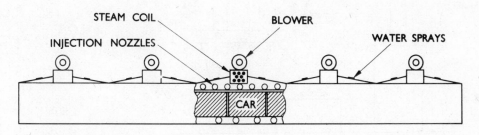

Fig. 9.13 Humidity dryer.

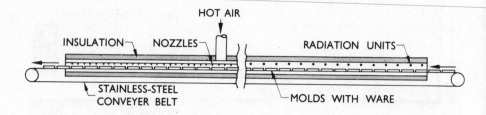

Fig. 9.14 Tableware dryer.

Tableware dryers. Ware coming from the jigger or roller machine rests on plaster bats, which are conveyed through the dryer. Here the ware encounters first infrared radiation and then high-velocity jets of hot air striking it at right angles (Fig. 9.14). In this way, the water content of the ware may be brought down from 25% at the start to the 5% needed for finishing in 10 or 15 minutes as compared to several hours with the old mangle dryer. Of course, the temperature of the molds must not exceed 70°C.

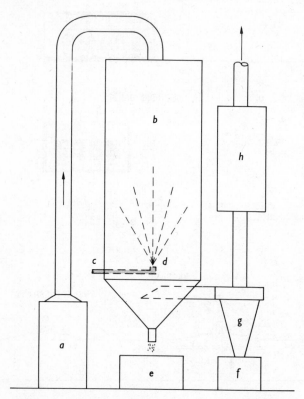

Fig. 9.15 One type of spray dryer: *a*, air heater; *b*, drying chamber; *c*, slip source; *d*, spray nozzle; *e*, dry particles; *f*, fines; *g*, air separator; *h*, dust collector.

Spray dryers. These dryers are much used in preparing the batch for dry-pressed fine ceramics, to convert a slip into a compressible powder. There are several types, but they all atomize the slip by passing it through spray nozzles or a centrifugal distributer and then allow the droplets to fall through heated air (Fig. 9.15). The product is fairly uniformly sized spheres, either solid or hollow.

Dielectric drying. In this method, the ware is passed between a pair of plates excited by a 10-megacycle source, which supplies heat uniformly through the ware by atomic vibration. The rate of temperature rise of the ware is:

$$\frac{2\pi \text{ (dielectric constant)} \times \text{(loss factor)} \times \text{(field strength)}^2 \times \text{(frequency)}}{\text{(specific gravity)} \times \text{(specific heat)}}$$

This method has been used for drying foundry cores, but, since only 15 to 20% of the electrical input goes into the ware as heat and since the equipnent is costly, the ceramic industry has not yet used it to any extent.

References

Sherwood, T. K., The Drying of Solids, *Ind. Eng. Chem.* **21**, 12, 1929; **21**, 976, 1929; **24**, 307, 1932

Macey, H. H., The Relative Safe Rates of Drying of Some Different Sizes and Shapes, *Trans. Brit. Ceramic Soc.* **38**, 464, 1939

Holdridge, D. A., The Effect of Moisture Content on the Strength of Unfired Clay Bodies, *Trans. Brit. Ceramic Soc.* **51**, 401, 1952

Pask, J. A., Measurement of Dry Strength of Clay Bodies, *J. Am. Ceramic Soc.* **36**, 313, 1953

Dollimore, D., and S. J. Gregg, The Effect of Water Absorbtion on Kaolinite Compacts, *Trans. Brit. Ceramic Soc.* **54**, 262, 1955

Woo, D., *et al.*, Drying Rates in Infrared Drying of Clay, *J. Am. Ceramic Soc.* **38**, 383, 1955

Moore, F., The Mechanism of Moisture Movement in Clays with Particular Reference to Drying: A Concise Review, *Trans. Brit. Ceramic Soc.* **60**, 517, 1961

Lancaster, B. W., and E. R. McCartney, Humidity Drying of Plastic Clay, *J. Am. Ceramic Soc.* **47**, 291, 1964

Ryan, W., Factors Influencing the Dry Strength of Clays and Bodies, *Trans. Brit. Ceramic Soc.* **64**, 275, 1965

Williamson, W. O., Strength of Dried Clay – A Review, *Bull. Am. Ceramic Soc.* **50**, 620, 1971

10

Firing ceramic ware

1. Introduction

Of all the steps in the process of producing ceramic articles, the firing is the most vital. Therefore we should learn all that we can about the reactions that take place at high temperatures. Of special interest is the equilibrium condition that is being approached, even though that condition is not always reached. Also, the rate at which the reaction progresses is often important.

2. Thermodynamics of Reactions

Energy states. The stable phase at any temperature level is the one having the lowest free energy. For example, low quartz and quartz glass both exist at room temperature. If each is dissolved in hydrofluoric acid and the heat of the solution measured, it will be found that the value for the glass is the higher. This shows that the glass has the higher free energy and is therefore the unstable phase.

Heats of formation. When two elements combine, they absorb or evolve energy ΔH, as shown for some of the oxides in Table 10.1. In general, if the heat energy evolved is large, the oxide has a high stability, as is the case for MgO or ZrO_2, but the fusion point is not exactly proportional to the heats of formation.

Vapor pressures. The stability of ceramic compounds is often indicated by the vapor pressure; the lower this value, the greater the stability. Values for a few oxides are given in Table 10.2.

Equilibrium conditons. The equilibrium between FeO and Fe_2O_2 is important for the color of glass. At high temperatures the equilibrium goes toward FeO and at higher oxygen pressures toward Fe_2O_3. The equilibrium under various conditions may be calculated from thermodynamic data. However, thermodynamics cannot compute the rate at which equilibrium is reached.

Table 10.1 Heats of formation
of some oxides

Oxide	ΔH^1 in cal/mol
MgO	-144,000
CaO	-152,000
Al_2O_3	-400,000
SiO_2	-206,000
ZrO_2	-259,000
BeO	-147,000
FeO	-65,000
NiO	-53,000
PbO	-53,000

[1]When ΔH is minus, an evolution of heat is indicated.

Table 10.2 Vapor pressures
of some oxides

Oxide	Pressure of O_2 in atmos at $1000^\circ K$
CaO	4×10^{-56}
BeO	4×10^{-54}
MgO	3×10^{-53}
Al_2O_3	4×10^{-46}
ZrO_2	3×10^{-47}
SiO_2	4×10^{-36}
ZnO	2×10^{-28}
SnO_2	1×10^{-20}
CoO	9×10^{-18}
NiO	3×10^{-18}
PbO	2×10^{-13}
CuO	2×10^{-11}

It is strongly recommended that the student of ceramics obtain a good
background in that part of physical chemistry dealing with thermodynamic
processes. There is space here for only a glance at the subject.

3. Phase Rule

The statement of the phase rule by Gibbs was one of the great contributions to science. On it is based the physical chemistry of the ceramic field. Therefore, the student should be familiar with it as a groundwork for his ceramic work. There is no space here to do more than touch on this subject. It should have been covered by previous courses in physical chemistry.

The phase rule may be stated as

No. of phases + No. of degrees of freedom = No. of components + 2,

or

$$P + V = C + 2. \tag{1}$$

A phase is a physically homogeneous but mechanically separable portion of a system. Examples of phases in a system are ice and water.

The degrees of freedom of a system are the number of independent variables, such as temperature, pressure, and concentration, which must be fixed to define the system completely.

The components are the smallest number of independently variable constituents by means of which the composition of each phase can be quantitatively expressed. For example, in the system Mg and O the components could be Mg, O; MgO, O; or MgO, Mg.

4. Equilibrium Diagrams

The equilibrium diagram is a graphic representation of the phase rule as applied to a particular system. The two variables are usually temperature and composition. These diagrams give a complete picture of the system under specific conditions and thus are most useful for reference.

Method of making equilibrium diagrams. In the case of many metals the diagram may be readily determined by heating the components together in definite proportions until only the liquid phase is present. This liquid is allowed to cool slowly with a careful measurement of its temperature. The cooling curve will have breaks at the start and end of crystallation that make it possible to construct the diagram of this one composition readily. By repeating with other compositions the whole diagram may be obtained.

In the case of ceramic materials the heat evolved in crystallization is often small and the possibility of undercooling is great, so another method must be found. This consists of heating the intimately mixed components to a given temperature until equilibrium is obtained and then quenching them to room temperature. A petrographic or x-ray examination will then give the kind and amount of phases present. This is repeated for other compositions and other temperatures until the diagram is completed. The complicated nature of the diagram, as well as the large number of runs needed, limits most work to two or three components.

Interpretation of equilibrium diagrams. In Figs. 10.1 to 10.5 are shown a number of simple phase diagrams. Figure 10.1 is a single-component diagram with pressure as the ordinate and temperature as the abscissa. At point *a*, for example, there are three phases, crystal *A*, liquid, and vapor. Using the phase rule, we get

$$P + V = C + 2$$
$$3 + V = 1 + 2$$
$$V = 0.$$

Thus *a* is an invariant point with no degrees of freedom. Along the line *ab*, there are two components, crystal *A* and vapor, so here there would be one degree of freedom, either pressure or temperature. In any of the fields there would be only one phase, giving two degrees of freedom, both pressure and temperature.

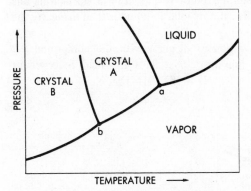

Fig. 10. 1 Single-component equilibrium diagram.

In ceramic systems the vapor phase is not generally of importance. Thus it is convenient to work at a fixed pressure of one atmosphere and have what is known as the condensed system. This permits us to use the so-called isobaric phase rule, $P + V = C + 1$.

In Fig. 10.2 there is shown a two-component diagram with one compound *AB* and two eutectics. In the same way as for the previous diagram it may be shown

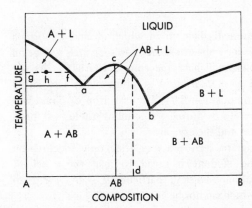

Fig. 10. 2 Two-component equilibrium diagram.

that the eutectic points a and b have three phases, liquid and two solids, and consequently

$$P + V = C + 1$$
$$3 + V = 2 + 1$$
$$V = 0.$$

If a composition AB cools, it will completely crystallize at point c and the solid phase will cool in this form. However, if a composition at d is cooled, the first crystals will come out at the liquidus curve, and the amount of crystals will increase as the temperature falls until the solidus line is reached. There the last of the liquid phase disappears.

The relative amounts of crystal and glass in a field may be determined graphically as follows. Take, for example, point h and draw a horizontal line intersecting the liquidus boundary. The ratio of solid to liquid is then hf/hg, by the so-called lever principle.

In Fig. 10.3 is shown a two-component diagram with the compound AB decomposing before it reaches the liquidus curve. This is called incongruent melting.

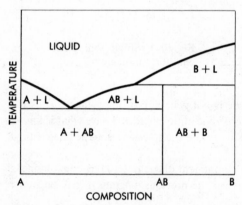

Fig. 10.3 Two-component equilibrium diagram with a compound AB melting incongruently.

Figure 10.4 shows a two-component diagram in which A and B form a continuous series of solid solutions. If a composition AB_2 cools, crystals will begin to form at the liquidus curve and these will have the composition AB_1. As the temperature drops the crystals will gain in B until at point c the last liquid disappears, and the crystal has the composition AB_2. The last drop of liquid will then have the composition L_1. Therefore, during the cooling there will be a constant interchange between the liquid and the crystals.

This discussion of the phase diagrams has been very brief, only touching on some of the important points. The student is urged to read the excellent presentation of this subject by Hall and Insley, particularly with regard to three-component diagrams, a subject which cannot be covered here.

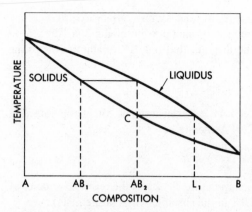

Fig. 10. 4 Equilibrium diagram with a continuous solid solution.

Example of an equilibrium diagram. As an example of the equilibrium diagram in the ceramic industry, the SiO_2-Al_2O_3 binary system is most appropriate. As shown in Fig. 10.5 there is one compound in the system, mullite, that melts incongruently at 1810°C. The eutectic lies close to the silica end. It would be expected from this· diagram that a brick with 70% alumina would contain some glassy phase on cooling down to the eutectic temperature of 1545°C, whereas a brick of 74% alumina would have no glass content below 1810°C. Actually it is true that bricks a little below the mullite composition deform under loads above 1500°C, while those a little higher in alumina stand up well to 1750°C.

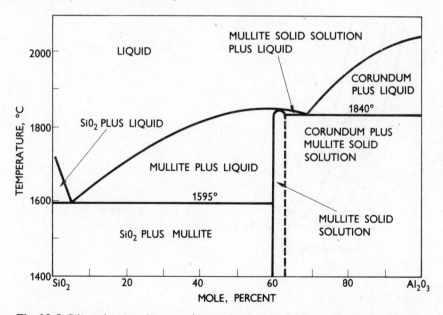

Fig. 10.5 Silica-alumina diagram. (Aramaki, S., and R. Roy, The Mullite-Corundum Boundry in the Systems MgO-SiO_2 and CaO-Al_2O_3-SiO_2. *J. Am. Ceram. Soc.* **42**, 644, 1959.)

In making silica brick, it has been found that a more refractory product may be made by washing the ganister to remove a small portion of alumina. The reason for this is made clear by noting on the diagram that only 1% of alumina drops the liquidus temperature by 15°C.

5. Reaction Rates

Chemical reactions in general proceed at a faster rate as the temperature is increased. This relation was expressed by Arrhenius as

$$\log\frac{k_{t2}}{k_{t1}} = A\,\frac{1}{T_1} - \frac{1}{T_2},$$

where k_{t1} is the reaction velocity at temperature T_1, k_{t2} is the reaction velocity at temperature T_2, and A is a constant (T is in the absolute temperature scale).

The reactions involved in firing a ceramic body follow this general law, so that a piece held at temperature for only one hour would require a higher temperature than a piece held for 10 hours to arrive at the same degree of maturity. For example, Fig. 10.6 shows the porosity of a vitrified body fired for various lengths of time and at different temperatures. Each tenfold increase in time corresponds to a 23°C decrease in temperature. This corresponds to the constant A in the Arrhenius equation of about 10,000, a value quite closely adhered to in many ceramic reactions.

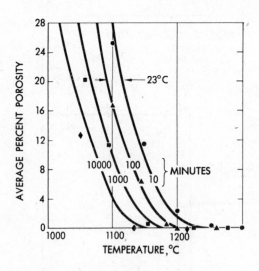

Fig. 10.6 Maturing conditions in a whiteware body.

6. Solid-State Reactions

Nearly all reactions in the traditional ceramic bodies were accompanied by some liquid that acted as a means of atomic transfer. More recently it has been believed possible to carry on reactions between two solids with no liquid phase. It seems now that such reactions may occur, but there is always the possibility that they may be sparked by a trace of impurity.

Single components. A solid-state reaction may occur in one component because of a crystal growth. This is due to the loosening of atomic bonds by thermal energy so that the free atoms or atom groups tend to fall on the larger crystal faces that have greater surface energy. Therefore, the small crystals grow smaller, and the large ones larger.

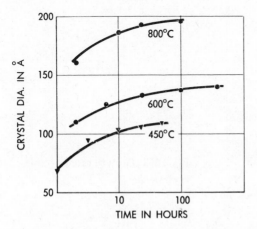

Fig. 10.7 Growth of MgO crystals from $MgCO_3$.

It should be kept in mind that not all reactions in ceramics follow the Arrhenius equation. For example, if MgO is formed by calcining $MgCO_3$ at various temperatures and for various times, the average size of the MgO crystal is as shown by the curves of Fig. 10.7. While the size of the initial crystal formed is dependent on temperature, the growth rate of that crystal is the same for all temperatures and decreases with elapsed time. If the growth rates, not only for MgO but also for other oxides, taken from the slope of the previous curves are now plotted against time, a straight line on a log-log plot will result (Fig. 10.8). This relation may be expressed by

$$\log R = \log R_o - n \log t$$

$$R = R_o / t^n = \frac{31}{t^{1.56}},$$

where R_o is the rate of growth in Ångstroms per second when the time is one hour, and n is the slope of the straight line. This illustrates the usefulness of a log-log plot in determining the value of an exponential.

Two components. If two components, A and B, are present, a slow diffusion process may gradually form the crystal AB. The greater the surface, due to fine

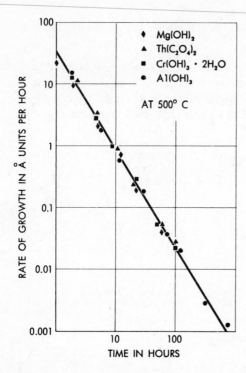

Fig. 10.8 Growth rate of oxide crystals.

comminution, the more rapidly the reaction goes forward. Also if the bonds are loosened by the passing of one crystal through a transformation, the reaction rate is increased.

Some examples of solid reactions. A single-component reaction may be illustrated by the growth of MgO crystals as shown in the previous section.

An example of a two-component reaction is the formation of spinel from MgO and Al_2O_3 at temperatures below the melting point. This reaction goes to completion rapidly by mutual diffusion if the two components are finely ground and intimately mixed.

7. Means for Measuring Thermochemical Changes

Shrinkage and expansion. All ceramic materials show volume changes in the kiln. Most of these are irreversible because the pores close up, but some of the polymorphic crystal changes are reversible.

At temperatures up to 1700°C, linear changes can best be indicated or recorded with the simple dilatometer of Fig. 10.9. Irreversible changes can be measured by heating the specimen to a series of temperatures, then cooling and measuring. Of course, results from dynamic tests with a rising temperature will be different from tests where the temperature is held at one point for some time.

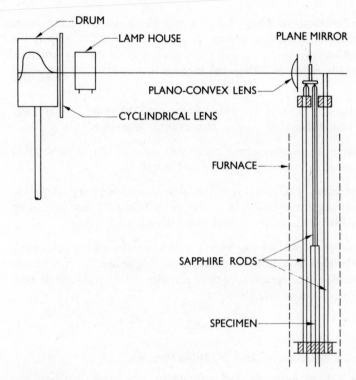

Fig. 10.9 Recording dilatometer. (Taken from F. H. Norton, *Fine Ceramics*, McGraw-Hill, New York, 1970.)

Porosity measurements. A knowledge of the porosity of a body is of importance for it serves as a measure of maturity and also allows the evaluation of a body for a specific purpose. The open pores in a fired body may be measured by the volume of air or water needed to fill the pores according to the following equation

$$P = 100 \, \frac{V_p}{V_b} = 100 \, \frac{W_s - W_d}{W_s - W_u},$$

where P is percent porosity, V_p the pore volume in cc, V_b the bulk volume in cc, W_s the sample weight saturated with water, W_d the sample weight dry, and W_u the sample weight saturated and immersed.

The total pore volume both closed and open may be found by

$$P = 100 \, \frac{V_p}{V_b} = 100 \, (1 - \frac{d}{\text{s.g.}}),$$

where d is the bulk density, and s.g. the true specific gravity.

The open pores may be measured for size as well as total volume by immersing the specimen in mercury under various pressures; the higher the pressure, the smaller the pore filled (see Chapter 23).

Heat evolution and absorption. When clay, for example, is raised in temperature alongside a neutral material, such as alumina, there will be a series of temperature differences between the two that gives much information about the clay structure. There is now available excellent differential thermal analysis (D.T.A.) equipment for such studies.

Weight loss. As in the case of shrinkage, weight loss may be determined dynamically or statically. Recording balances are now available for the latter case.

Petrographic examination. In cases where the crystals are over 10 to 15 microns in size the microscope will determine species, size, and morphology.

X-ray diffraction methods. This tool is very helpful in determining a crystal species, and by line broadening the crystal size in the range below 0.1 micron can be measured. Electron-diffraction methods work down to even smaller sizes.

The electron microscope. This is a very powerful tool for studying fine particles and crystals in a body or glass. Resolution approaches ten Ångstroms. The microprobe is useful in determining the composition of small particles and the scanning electron microscope gives a picture of surface character.

8. Thermochemical Reactions in Clays and Other Materials

Kaolin. As kaolin is both pure and well crystallized, it naturally has been the subject of most of the thermochemical studies in the clay field. On first heating, the mineral in kaolin called kaolinite loses the layers of adsorbed water which were in equilibrium with the atmosphere, but it is doubtful whether the innermost layer of molecular water is completely gone until relatively high temperatures are reached.

The first crystalline change in kaolinite, dehydroxylation, occurs at 450°C, when a weight loss signals the break-up to form meta-kaolin ($Al_2O_3 \cdot SiO_2$). In the unit cell of kaolinite there are 8 (OH) which break down into 4 0^{--} plus 4 H_2O. The O^{--} should be able to hold the original structure in some sort of order. Although x-ray diffraction methods show no definite crystal structure at this point, electron diffraction patterns are much like kaolinite, which indicates some of the original structure is left in a very fine-grained form. The heat required to complete the dehydroxylation is considerable, 35,000 cal (g.mole)$^{-1}$.

The next major change comes at around 925°C, corresponding to the strong exothermic peak at 980°C shown on the D.T.A. curve of Fig. 10.10. For many years it was not clear exactly what happened to the structure at this point, but recently Brindley and Nakahira have shed light on it. The meta-kaolin suddenly crystallizes into a spinel structure oriented in the same way as the original kaolinite with rejection of silica. This spinel quickly breaks down into mullite, still with the same orientation, at the same time rejecting more silica which forms glass.

At higher temperatures the mullite crystals grow larger and at 1200°C there is a sudden growth of secondary mullite, giving a small exothermic peak on the

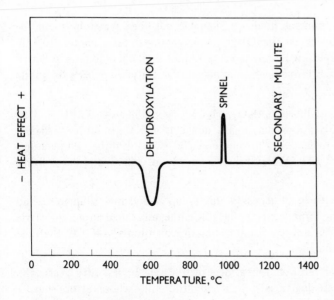

Fig. 10.10 Differential thermal
analysis curve for kaolin.

D.T.A. curve. Apparently cristobalite does not form from pure kaolin, although it
does in less pure clays.

The reader is referred to the excellent discussion of the topotactic changes
occurring in kaolinite by W. D. Johns (1965), who clearly describes the
transformations.

Other clays. Clays that contain impurities or are poorly crystallized show less
positive thermal effects than does kaolinite. The common accessory minerals
decompose as shown in Table 10.3.

Table 10.3   Decomposition of clay impurities

Reaction	Breakdown temp.,$^{\circ}$C
$FeS_2 + O_2 \rightarrow FeS + SO_2$	350–450
$4FeS + 7O_2 \rightarrow 2\ Fe_2O_3 + 4SO_2$	500–800
$Fe_2(SO_4)_2 \rightarrow Fe_2O_3 + 3SO_3$	560–775
$C + O_2 \rightarrow CO_2$	350→
$S + O_2 \rightarrow SO_2$	250–920
$CaCO_3 \rightarrow CaO + CO_2$	600–1050
$MgCO_3 \rightarrow MgO + CO_2$	400–900
$FeCO_3 + 3O_2 \rightarrow 2Fe_2O_3 + 4CO_2$	800→
$CaSO_4 \rightarrow CaO + SO_3$	1250–1300

Montmorillonite. This mineral holds considerable adsorbed water both on the surface and between the layers. The $(OH)^-$ groups are driven off at about 450°C to form an amorphous appearing material. There is a small exothermic peak at 800°C where a spinel structure is formed, which at higher temperatures converts to mullite and glass.

Hydrated alumina. When gibbsite $(Al_2O_3 \cdot 3H_2O)$ is heated in air to 150°C, it alters to boehmite $(Al_2O_3 \cdot 3H_2O)$, which breaks down at 425°C to form $\gamma-Al_2O_3$. Bayerite $(Al_2O_3 \cdot 3H_2D)$ changes to boehmite at 200°C and at higher temperatures to $\alpha-Al_2O_3$. Diaspore $(Al_2O_3 \cdot H_2O)$ breaks down at 350°C and later forms $\gamma-Al_2O_3$.

Kyanite minerals. The three minerals in this group — kyanite, sillimanite, and andalusite — all have the formula $Al_2O_3 \cdot SiO_2$. Kyanite, most used at present, starts to decompose at 1100°C and reaches a maximum expansion at 1400°C, the final product being mullite and glass.

Silica. This material has received a great deal of study in the last fifty years. The usual minerals of SiO_2 are quartz, cristobalite, and tridymite, whose relationships at atmospheric pressure can best be shown in the diagram of Fig. 10.11.

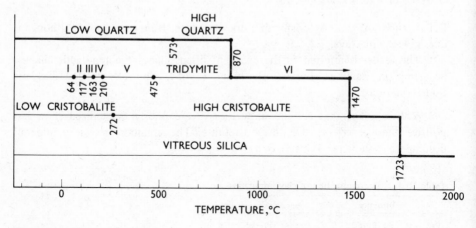

Fig. 10.11 Phases of silica.

Sosman reports some high-pressure forms of SiO_2. Keatite occurs at pressures between 0.8 and 1.3 kb at temperatures of 400 to 550°C. Coesite occurs between 15 and 40 kb at temperatures of 300 to 1700°C. The mineral Stishovite with a rutile structure exists at some 160 kb and temperatures between 1200 and 1400°C.

Cristobalite can be formed from quartz at elevated temperatures in a sluggish inversion, sometimes hastened by addition of impurity atoms.

Tridymite has a more open structure than quartz and this accounts for its low density. The fact that it is never found without impurity atoms now leads some to believe that it should not be called a silica mineral.

There are hydrated forms of silica, the important one in ceramics being flint, a cryptocrystalline quartz with 1% water in the structures. On heating in the range of 300°C the structure becomes cracked due to steam pressure. At 1000°C the last water is gone and at 1200°C the quartz inverts to cristobalite.

Table 10.4 shows the densities of the various forms of silica.

Table 10.4 Density for forms of silica

Mineral	Density, $g \cdot cm^{-3}$	
	Low form	High form
Quartz	2. 65	2. 63
Tridymite	2. 27	2. 26
Cristobalite	2. 33	2. 29
Quartz glass	2. 21	
Flint	2. 61	

Feldspars. At 1100°C soda feldspars melt but potash feldspars decompose into leucite which does not completely melt until 1500°C is reached. This relationship is shown in the diagram of Fig. 4.1.

Talc. This mineral, $(OH)_2$ Mg_3 $(Si_2O_5)_2$, loses water at around 900°C to form protoenstatite and glass and at around 1300°C clinoenstatite $(2MgO \cdot SiO_2)$, cristobalite, and glass result. It should be kept in mind that the talc used in ceramics is often a mixture of a number of minerals, so the picture of the heat effects is not at all clear.

Mica. The platelike mineral, muscovite, breaks up at around 1000°C, whereas the man-made fluorine mica (phlogopite) is stable up to 1200°C.

Carbonates. Both lime and magnesia start to calcine at a definite temperature but the rate of CO_2 evolution depends on the size of the original crystals. The action starts at the surface of the crystals and works inward as the CO_2 diffuses outward. After the CO_2 is lost the atoms form nuclei of the oxide and these grow rapidly at first and then at an ever-decreasing rate. At any one time, the rate is increased by elevating the temperature.

9. Effect of Heat on Ceramic Bodies

Triaxial bodies. The average triaxial body is about 50% clay, 25% quartz, and 25% feldspar. The feldspar and quartz will have a particle size averaging 40 microns, the kaolin 2 microns, and the ball clays 0.3 microns. By chance this combination forms a system suitable for close packing, allowing the green body to have a porosity of 30 to 40%. Also, no matter what forming method is used, there will be some preferred orientation of the particles, particularly of the kaolin.

We do not have a complete picture of the life history of a triaxial body on firing, but the reactions summarized in Table 10.5 give the probable events. The usual triaxial body contains little or no cristobalite after firing but some bodies at high temperatures show considerable inversion of the quartz. Properly matured bodies have not over 0.1% of open pores but the closed pores amount to 3 to 5% unless the body is fired under special conditions, such as in a helium atmosphere or a vacuum. Figure 10.12 is an electron microphotograph of a sanitary-ware body showing much detail.

Table 10.5 Life history of a triaxial body

Temperature, ^{o}C	Reactions
Up to 100	Loss of moisture
100–200	Removal of adsorbed water
450	Dehydroxylation
500	Oxydation of organic matter
573	Quartz inversion to high form. Little overall volume damage
980	Spinel forms from clay. Start of shrinkage
1000	Primary mullite forms
1050–1100	Glass forms from feldspar, mullite grows, shrinkage continues.
1200	More glass, mullite grows, pores closing, some quartz solution
1250	60% glass, 21% mullite, 19% quartz, pores at minimum

In Fig. 10.13 is shown the shrinkage and porosity curve of a typical vitreous triaxial body (high-tension electrical porcelain) with a maturing temperature of 1250°C when held for 6 hours. It will be seen that overfiring causes the closed pores to expand, thus opening up the structure.

Translucency, an important property of fine vitreous ware, is increased when pores in the body are at a minimum and the index of refraction of the glass phase approaches that of the crystalline phases. As there is a considerable difference between the index of refraction of quartz and mullite, a high translucency cannot

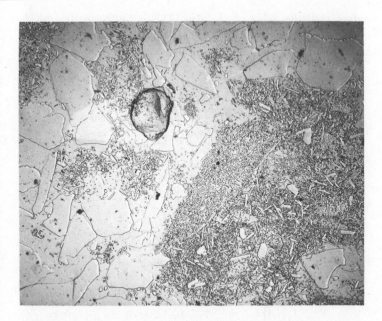

Fig. 10.12 Cone 13 sanitary-ware body polished and etched, then electron-microphotographed at a magnification of 4000 ×. The faintly outlined quartz particles show only slight solution. The fine scaly material is primary mullite from the clay and the needles are secondary mullite.

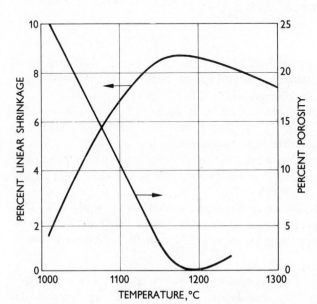

Fig. 10.13 Typical shrinkage and porosity curve for a high-tension electrical porcelain body.

be reached no matter what the index of the glass. This is shown in Fig. 10.14, which gives the results when mixtures of finely ground quartz and mullite saturated with liquids of varying index of refraction were measured for transmission.

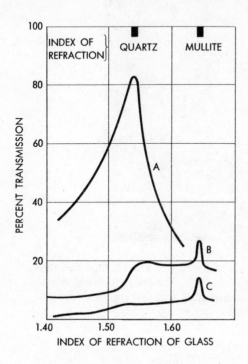

Fig. 10.14 Showing the influence of the glass phase on translucency; (A) 40% quartz, 60% glass; (B) 20% quartz, 20% mullite, 60% glass; (C) 30% mullite 70% glass.

Other fine-grain bodies. Bodies with low talc additions form a fluid glass phase which results in a short vitrification range. High talc bodies are in the clinoenstatite field but there is little information available on the life history during firing.

Bone-china bodies have been studied by St. Pierre (1955), who showed that the matured structure was mainly β-tricalcium phosphate and glass. This combination is responsible for the high translucency.

Lithia-containing bodies have been studied thoroughly by Hummel. Their low thermal expansion is due to the formation of β-spodumene solid solution and β-eucryptite solid solution.

Cordierite bodies for low-expansion ware are found in the $MgO \cdot Al_2O_3 \cdot SiO_2$ system and consist largely of the mineral cordierite ($Mg_2Al_2Si_5O_{15}$). Bodies with the lowest expansion consist of a mass of very fine cordierite crystals with little or no glass.

Heavy refractory bodies. Fire-clay refractories usually consist of a coarse grog bonded with clay. Firing properties depend on the type of clay used and the

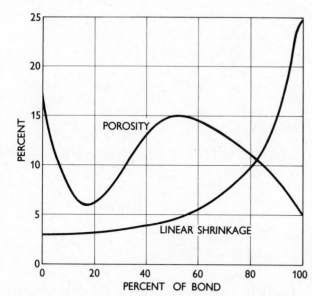

Fig. 10.15 Shrinkage and porosity for mixtures of grog and kaolin.

amount. The shrinkage curve in Fig. 10.15 is for bodies with the whole range of clay-to-grog ratio. The grog alone has a low shrinkage which does not change until the pores are filled with clay, after which the shrinkage increases as clay partitions are built up between the grog grains.

Silica bricks, of ganister with about 3% lime added as a mineralizer, expand in firing by 3 to 5% linearly as some of the quartz inverts to cristobalite and tridymite.

Basic bricks consist of magnesite, chromite, or a mixture of the two. The magnesite converts to periclase between 1700 and 1800°C. Because the impurities in the chromite form glass, it is attempted to get direct bonding between the crystalline constituents by high firing.

10. Solid-State Sintering

Introduction. In the last twenty years there has been much effort to understand the mechanism of sintering crystalline material when no glass phase is present. Advantage has been taken of the theoretical background built up in this field to produce articles with previously unobtainable properties.

Mechanism of solid-state sintering. A simple model for sintering studies is an assembly of even-sized spheres (Fig. 10.16(a)). When this assembly is heated there is a welding together at the contact points and a neck is built up (Fig. 10.16(b)). As heating continues the pores become closed and spherical as in Fig. 10.16(c), all these changes being accompanied by an overall volume shrinkage. Further treatment causes these pores to diminish in size (Fig. 10.16(d)) and finally disappear.

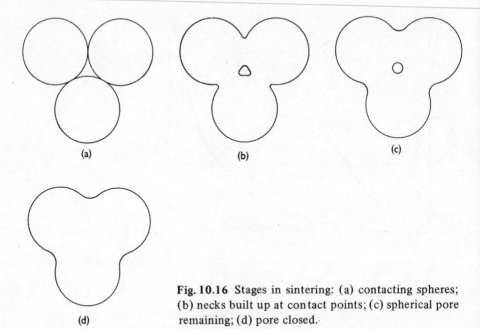

Fig. 10.16 Stages in sintering: (a) contacting spheres;
(b) necks built up at contact points; (c) spherical pore
remaining; (d) pore closed.

It is evident that the sintering described in the previous paragraph entails a movement of material. This may be a solid flow, a surface flow, or a transfer in a gaseous phase. Also, the gas trapped in the closed pores must escape either by diffusion along grain boundaries or through the crystal lattice. There is not space here to go deeply into this mechanism, but the references at the end of the chapter will throw much light on the subject.

Sintered alumina. This oxide, sintered to high density, is much used for electrical, refractory, and abrasion-resistant purposes. The raw material, containing 96 to 99% Al_2O_3, is in the form of fine crystals of about 0.1 micron diameter. In order to reach high densities a sintering aid is needed, usually 0.05% MgO. This prevents rapid crystal growth by concentrating in the grain boundaries, and seems to open up the grain boundaries to allow diffusion of gases from the pores. It has also been found that lower porosity can be obtained by firing in gases of high diffusion rate

Table 10.6 Methods of firing alumina

Material	Conditions of firing	Density, %
Pure Al_2O_3	Air	95
Pure Al_2O_3	O_2, H_2, or vacuum	95
Al_2O_3 + 0.05% MgO	Air	98
Al_2O_3 + 0.05% MgO	O_2, H_2, or vacuum	100

Fig. 10.17 Commercial sintered alumina, 250 X. (Avco Corporation.)

such as O_2 or H_2. This is shown in Table 10.6 from the work of Coble. The maturing temperature is between 1800 and 1900°C. Figure 10.17 shows a microsection of pure commercial sintered alumina with a 10-micron grain size.

Sintered beryllia. This material is used for heat sinks because of its high thermal conductivity. It is sintered like alumina with sintering aids of 0.3% talc or magnesia.

Sintered magnesia. The pure MgO is fired to 1900 or 1950°C to produce a dense, translucent ware.

11. Hot Pressing Pure Oxides

In some cases it is necessary to form nonporous bodies of high-purity oxides where sintering aids are not permissible. By using moderate temperatures combined with high pressures, rapid grain growth is inhibited and the structures shown in Fig. 10.18 may be obtained.

Pressing is usually carried out in graphite molds at 4000 to 10,000 psi and temperatures from 1100 to 1400°C. At present, hot pressing is a high-cost operation and boron carbide is the only material made commercially in this way in any quantity.

12. Setting Methods

Heavy clay products. Brick are uniform and compact with reasonably good strength at firing temperatures, so they can be piled to heights of 10 to 20 feet in a checkerwork structure (Fig. 10.19). Sewer pipe are fired on end, often supported by a short, detached section of the same pipe to prevent shrinkage cracks (Fig. 10.20).

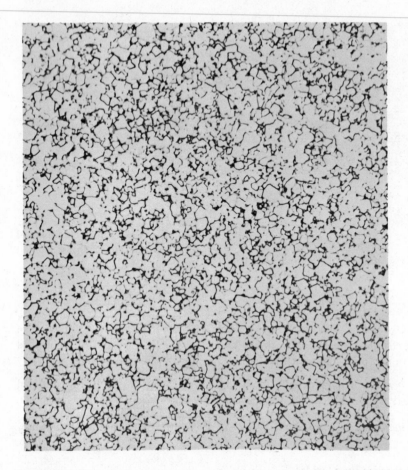

Fig. 10.18 Pure dense alumina, hot-pressed 99.95% Al_2O_3, 1000 X. (Avco Corporation.)

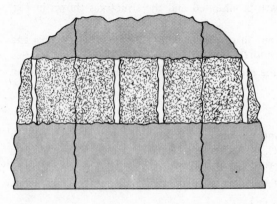

Fig. 10.19 A setting of brick in the kiln.

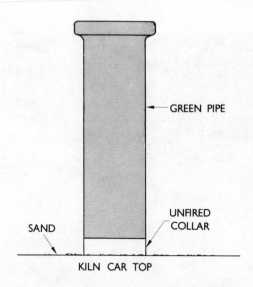

GREEN PIPE

UNFIRED COLLAR

SAND

KILN CAR TOP

Fig. 10.20 Method of setting sewer pipe.

Refractories. Brick in this classification are set in piles like building brick but, since the firing temperature is much higher, the piling must usually be lower. General practice is as follows:

Silica brick	40—60 courses in height
Fireclay brick	30—40 courses in height
Basic brick	1— 3 courses in height
Kaolin brick	1— 2 courses in height

However, in tunnel kilns, the setting height is seldom over 20 courses. Special refractory shapes may be set on shelves (Fig. 10.21).

Tile. Once-fired wall tiles are often placed in setters (Fig. 10.22) or they may be placed on refractory plates and sent through the kiln on rollers (Fig. 10.23). Floor tiles are packed in small saggers.

Tableware. Earthenware is biscuit-fired in bungs and glost-fired in open setters (Fig. 10.24). Vitreous ware, such as hotel ware, is biscuit-fired in bungs with a refractory powder between it to prevent sagging and glost-fired in saggers of open setters. Thin ware, like bone china, is biscuit-fired, placed on individual refractory setters coated with an alumina wash to prevent sticking.

High-tension electrical porcelain. Large pieces are set on end with an unfired setter as a base to allow diametrical shrinkage without cracking.

Electronic parts and special refractories. Large tubes are set on end on an unfired base and slender tubes are hung from the top so that gravity forces keep them straight. Alumina substrates are fired in small piles on zircon plates and radomes are set on large, unfired setters.

Fig. 10.21 Setting of special refractory shapes. (Babcock and Wilcox Co.)

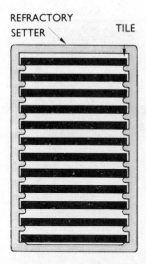

Fig. 10.22 A setting for once-fired tile.

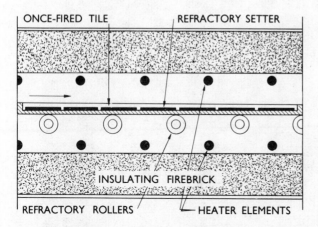

Fig. 10.23 Tile on bats in roller kiln.

Artware. One of the problems in making large pieces of vitreous artware is to hold that roundness of the rim during firing. This is often solved by placing setters in the opening. Sometimes these are fired setters cut at such an angle that they slide upward as shrinkage occurs (Fig. 10.25(a)) or they may be unfired setters supporting the rim (Fig. 10.25(b)).

Fig. 10.24 Glost setting of hotel china. (Swindell-Dressler Corp.)

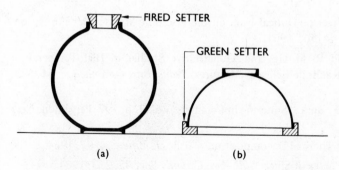

Fig. 10.25 Setting for large hollowware.

13. Finishing Fired Ware

Some fired ceramic products are given finishing operations. Low-tension porcelain insulators are tumbled in walnut shells to remove fins and rough edges. Hotel china bisque ware is tumbled or vibrated with pebbles or small porcelain cylinders to remove any roughness. Fine-grained sintered oxides are lapped and polished with diamond tools. Holes are drilled with diamond tools or in some cases with gas lasers.

Electronic components are sometimes metallized so they can be joined to metal parts. One process is as follows:

Heat to 1000°C and cool

Apply molybdenum-manganese alloy in powder form by roller, silk screen, or spray in a thin layer

Fire in a reducing atmosphere to 1475°C and cool

Electroplate with nickel, copper, or gold

Braze

Glost ware has pin marks ground off and often the foot is smoothed off.

References

Goodman, G., Relation of Microstructure to Translucency of Porcelain Bodies, *J. Am. Ceramic Soc.* **23**, 66, 1950

German, W. L., Observations on the Properties of Pottery Fired on Rapid Schedules in Some Types of Tunnel Ovens, *Trans. Brit. Ceramic Soc.* **51**, 198, 1952

St. Pierre, P. D. S., Constitution of Bone China: I, *J. Am. Ceramic Soc.* **27**, 243, 1954; II, *J. Am. Ceramic Soc.* **38**, 217, 1955

Brindley, G. W. and M. Nakahira, Kinetics of Dehydroxylation of Kaolinite and Halloysite, *J. Ceram. Soc.* **40**, 346, 1957

Comer, J. J., New Electron-Optical Data on the Kaolinite-Mullite Transformation, *J. Am. Ceramic Soc.* **44**, 561 1961

Brindley, G. W., and R. M. Ougland, Quantitative Studies of High-Temperature Reaction of Quartz-Kaolinite-Feldspar Mixtures, *Trans. Brit. Ceramic Soc.* **61**, 599, 1962

Coble, R. L., and J. E. Burke, Sintering in Ceramics, Vol. 3, p. 197, Pergamon, New York, 1963

Koltermann, M., The Thermal Decomposition of Talc, *J. Mineral* **4**, 97, 1964

Sosman, R. B., The Phases of Silica, *Bull. Am. Ceramic Soc.* **43**, 213, 1964

Weber, J. N., and R. Roy, Dehydroxylation of Kaolinite, Dickite and Halloysite, *Am. Mineral* **50**, 1038, 1965

Johns, W. D., A Review of Topotactic Development of High Temperature Phases from Two-Layer Silicates, *Bull. Am. Ceramic Soc.* **44**, 682, 1965

Callister, W. D., *et al.*, Thermal Decomposition Kinetics of Boehmite, *J. Am. Ceramic Soc.* **49**, 419, 1966

Dinsdale, A., and W. T. Wilkinson, Properties of Whiteware Bodies in Relation to Size of Constituent Particles, *Trans. Brit. Ceramic Soc.* **65**, 391, 1966

Gibbs, G. V., The Polymorphism of Cordierite, *Am. Mineral* **51**, 1068, 1966

Bruce, R. H., and W. T. Wilkinson, Fillers for Whiteware Bodies, *Trans. Brit. Ceramic Soc.* **65**, 233, 1966

Ruddlesden, S. N., Application of the Electronprobe Microanalyser to Ceramics, *Trans. Brit. Ceramic Soc.* **66**, 587, 1967

Fulrath, R. M., and J. A. Pask (eds.), *Ceramic Microstructures*, Wiley, New York, 1968

Prabhakaram, P., Exchangeable Cations and the High Temperature Reactions of Kaolinite, *Trans. Brit. Ceramic Soc.* **67**, 105, 1968

Carruthers, T. G., and B. Scott, Reactive Hot-pressing of Kaolinites, *Trans. Brit. Ceramic Soc.* **67**, 185, 1968

Poch, W., The Sintering of Fine-Particle Oxides under High Pressure, in *Science of Ceramics*, Vol. 4, British Ceramic Society, Stoke-on-Trent, 1968

Guillatt, I. F., and N. H. Brett, The Determination of the Crystallite Size of Ceramic Powders by X-ray Line-broadening Techniques. *J. Brit. Ceramic Soc.* **6**, 56, 1969

Johnson, H. B., and F. Kessler, Kaolinite Dehydroxylation Kinetics, *J. Am. Ceramic Soc.* **52**, 199, 1969

Manual on Use of Thermocouples in Temperature Measurement, A.S.T.M. Tech. Pub. No. 470, New York, 1970

Nicholson, P. S., and W. A. Ross, Kinetics of Oxidation of Natural Organic Materials in Clays, *J. Am. Ceramic Soc.* **53**, 154, 1970

Walker, E. G., and D. A. Holdridge, Vitrification and Expansile Characteristics of Feldspars and Feldspathic Fluxes, *Trans. Brit. Ceramic Soc.* **69**, 21, 1970

Budworth, D. W., Theory of Pore Closure During Sintering, *Trans. Brit. Ceramic Soc.* **69**, 29, 1970

11

Kilns

1. Introduction

The modern kiln has evolved through thousands of years of changes, each designed to better adapt it to its purpose of firing ceramic ware consistently and economically. In the last few decades the kiln has become more specialized to fit the needs of many new products. Modern kilns may be classified as follows:

> I. *Periodic kilns*
> Stationary
> Elevator
> Shuttle
> Scove
>
> II. *Continuous kilns*
> Rotary
> Tunnel
> Chamber (not used in the United States)

2. Periodic Kilns

Stationary kilns. In these traditional kilns, the ware is placed inside, the whole is heated to the maturing temperature, and then the whole is cooled down and the ware drawn. These kilns are relatively low in cost and flexible in use, but have a high fuel consumption, long firing time, and require a high cost of labor.

Some heavy clay products and refractories are fired in large round or rectangular down-draft kilns. The more recent kilns of this type have been improved in efficiency and rate of firing by the use of insulating firebrick linings.

Elevator kilns. Recently there has been increasing use of elevator kilns (Fig. 11.1), which are raised above the charge for setting and drawing. Burners in the corners

Fig. 11.1 An elevator kiln for firing high-alumina refractories (Bickley Furnaces, Inc.).

give a rotary flow that produces excellent temperature uniformity and the capability of reaching 1800°C. Another kiln of this type, with thin walls of insulating firebrick and Kanthal heating elements distributed over walls, floor, and roof, can be used up to temperatures of 1250°C.

Shuttle kilns. A kiln of this type (Fig. 11.2) has a moving car bottom that allows setting and drawing the charge in the open.

Fig. 11.2   A large shuttle kiln (Bickley Furnaces, Inc.).

All of these modern periodic kilns are characterized by very thin walls (4½ to 9 in.), usually of insulating firebrick. This means that the heat capacity is small and rapid cycles can be followed. Even large kilns of this type can be fired every 24 hours. This results in savings in fuel or electricity.

Scove kilns. These were the earliest kilns used by brickmakers, and even now are used by some brickyards. The green bricks are piled in a checkerwork fashion in huge rectangular piles with fire holes left along the base. The sides are plastered with soft clay to form an impermeable envelope (scoving). The kiln is burned off, formerly with wood, but now with oil. The temperature distribution is not very good, but any underfired brick can be refired. Some improvements consist of mechanized setting and drawing and portable scoving of insulating firebrick panels.

3. Continuous Kilns

Introduction. These kilns are characterized by a constant temperature pattern through which the ware passes at an even rate. This has the following advantages:

1. Waste heat is used to preheat ware
2. Cooling ware preheats combustion gases

> fuel economy

3. Each portion of the kiln is at constant temperature

> long structure life

4. Setting and drawing is conveniently located

> allows mechanization

On the other hand the continuous kiln is not flexible; if it is slowed down, efficiency falls off and shutdowns are costly. It must, therefore, be geared closely to production.

Rotary kilns. These kilns are tubular structures with a nearly horizontal axis about which they revolve slowly. They are used for firing cement clinker and grog for refractories, burning magnesite and lime, as well as drying many products.

Direct-fired tunnel kiln. This is the kiln most used in the industry at present, handling a great variety of products. A side view is shown in Fig. 11.3. It will be seen that the ware moves through the tunnel in one direction while the gases flow in the opposite way. This makes it possible for the combustion air to be preheated by the cooling ware and cool entering ware to be preheated by the combustion gases, in this way achieving an excellent recuperative action. The burners spread along a section in the center of the tunnel provide heat for the high-temperature zone.

Figure 11.4 is a cross section of a typical tunnel kiln. It will be seen that the charge fits the cross section of the tunnel closely, to force the moving gas through the ware. The kiln car runs on tracks with a sand seal to prevent escape of hot gases. As each part of the kiln remains at a constant temperature, the brickwork is under ideal conditions and there are records of kilns running continuously for 25

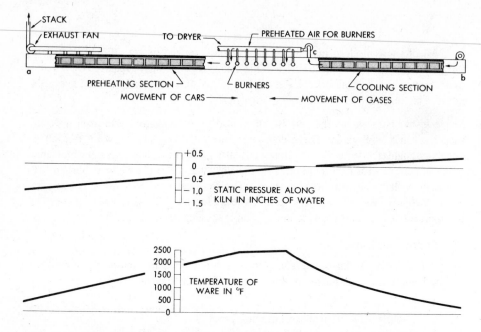

Fig. 11. 3 Side view of a direct-fired tunnel kiln.

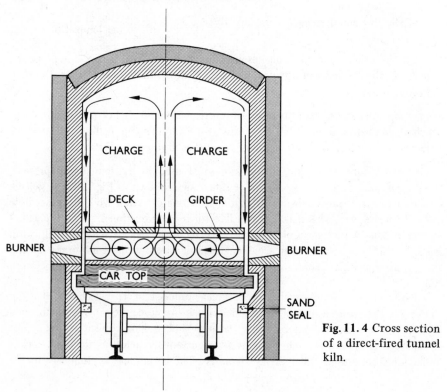

Fig. 11. 4 Cross section
of a direct-fired tunnel
kiln.

years. The car tops are a different story, since they are heated and cooled at each trip. Many methods of construction have been tried to increase their life. The kilns generally have an air lock at each end to control the gas flow better but kilns are running quite satisfactorily with one or even no air lock.

The cars are usually propelled through the kiln by a hydraulic pusher. Since this means that the train of cars stops completely after each stroke of the ram, it is desirable not to have the burners impinge directly on the charge or there will be local overheating. The rate of travel through the kiln varies a good deal but one-half to two 6-ft cars per hour would be normal.

The kiln is controlled by thermocouples (or radiation pyrometers for higher temperatures), which are set up to regulate the fuel supply. In addition test specimens or pyrometric cones are sent through to give a check.

Small tunnel kilns may push the ware through on refractory slabs instead of cars, with the advantage that heat may be applied beneath the slab as well as on top, allowing more even heating. Other kilns carry the ware through on a stainless-steel belt at lower temperatures and roller kilns, as in Fig. 10.23, are coming into use for firing tile.

Muffle tunnel kiln. This type of kiln, originated by Phillip Dressler, has been used for glost ware, but is gradually being replaced by the direct-fired kiln. The cars of ware pass down the tunnel and are heated by radiation and convection streams from muffles fired by longitudinal burners. This protects glost ware from direct contact with the combustion gases.

Controlled-atmosphere tunnel kiln. These kilns are a rather recent development to take care of firing electronic parts such as ferromagnetics. The kiln is generally small with air locks and gas generators to provide the correct atmosphere. The heat is supplied by electric elements, with very close temperature control. In some cases the ware is pushed through the kiln on slabs and in other cases it goes through on rollers. A photograph of such a kiln is shown in Fig. 11.5.

4. Kiln Efficiency

The maker of high-cost products such as electronic parts cares less about efficiency than he does about temperature and atmospheric uniformity, whereas the maker of heavy products is vitally interested in this problem, since fuel is a considerable portion of his cost.

The kiln efficiency has usually been defined as

$$\text{percent efficiency} = \frac{\text{heat to bring charge to maturing temperature}}{\text{heat delivered by fuel}} \times 100.$$

In the case of the periodic kiln the heat balance may be as shown in Fig. 11.6, where the efficiency is 20%. However, in the recuperative kiln the heat balance is as

Fig. 11.5 A small tunnel kiln for controlled-atmosphere firing (Harper Electric Furnace Corp.).

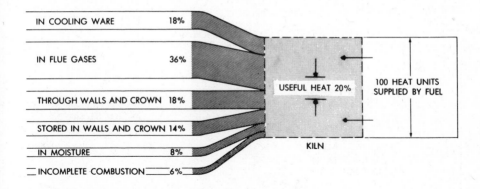

Fig. 11.6 Heat balance in a periodic kiln.

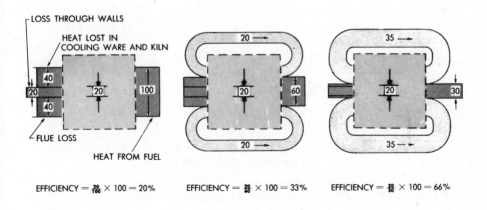

Fig. 11.7 The effect of increasing recuperative effect on calculated efficiency.

indicated in Fig. 11.7, showing a considerable increase in efficiency. It is quite possible, by using the above equation, to obtain greater than 100% efficiency, which suggests that such a definition is not satisfactory for recuperative kilns. It is better to use a figure of heat units from fuel for each pound of ware.

Figure 11.8 gives some values of heat units required to fire different types of ware in various kilns.

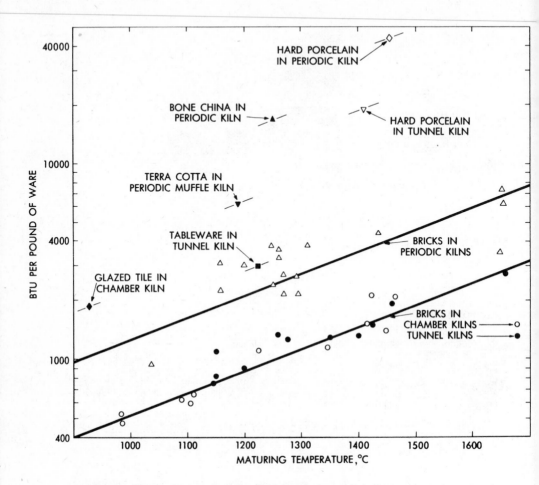

Fig. 11.8 Heat units required for firing in the ceramic industry.

References

McFadden, C. A., and G. B. Remmey, The Jet Burner — A New Concept in Fast Precision Firing, *Bull. Am. Ceramic Soc.* **41**, 160, 1962

Allen A. C., Roller-Hearth Kiln Fires in 20 Minutes, *Ceramic Ind.* **84**, 5, 50, 1965

Tatnall, R. F., Envelope Kiln Solves Space Problem, *Ceramic Age* **82**, 5, 22, 1966

Buchknemer, H., Europe Reports Progress in Whiteware Fast Firing, *Ceramic Ind.* **87**, 4, 62, 1955

Reed, R. J., and J. H. Lanzdorf, Firing and Controlling Kilns above 3000°F, *Bull. Am. Ceramic Soc.* **45**, 590, 1966

Buchnaner, R., and F. Petri, Kilns for Sintering Soft Ferrites, *Ceramic Age* **83**, 12, 32, 1968

12

The glassy state

1. Introduction

In Chapter 2 it was stated that solid matter exists either in the crystalline state, where the atoms are in an orderly array, or in the amorphous state, where the atoms form a random three-dimensional network. It is the purpose of this chapter to examine the latter state of matter with regard to its nature, its stability, and its properties. It should always be kept in mind that the important characteristic of glass is its transparency.

2. Constitution of Glass

Glass has been defined in simple terms as an undercooled liquid of very high viscosity, but a more precise definition is that given by Morey: "A glass is an inorganic substance in a condition which is continuous with and analogous to the liquid state of that substance, but which as the result of having been cooled from a fused condition has attained so high a degree of viscosity as to be for all practical purposes rigid." For many years the nature of glass was not understood. Only when x-ray diffraction techniques were brought to bear on this problem by Warren and his co-workers was it possible to obtain a clear picture, and then only for the simple glasses.

Glass network. The x-ray spectrum of a crystal has many sharp lines, but glass produces only a few very diffuse bands, showing that little order is present. The first diffuse band for silica glass represents the Si-O distance, the second, the O-O distance; after this the randomness of the structure precludes many other bands. It should be kept in mind that not all the Si-O distances are exactly alike because of the varying angles of the bonds and nonrepeating neighbors. This accounts for the great width of the bands.

 The rather involved method of translating x-ray diagrams into the structure cannot be touched on here, but the student who is interested may find this information in the references. It is sufficient to say that the final conclusion may be illustrated by the rather conventionalised two-dimensional drawings in Fig. 2.1(b). The crystal has a uniform network with a unit repeated time after time in all directions, whereas the glass structure is a random network with only the silicon-oxygen tetrahedron a nearly invariable unit. In this random network the continuous Si-O network can be interrupted by other atoms such as sodium and calcium. Because of the randomness it is not necessary to have stoichiometric proportions of these additional atoms as it is in a simple crystal; on the contrary, they may occur over a considerable range as the holes are gradually filled.

On the other hand, there is no certainty that the random structure is uniform throughout; it is possible in a sodium silicate glass, for example, that some small volumes might have the composition SiO_2 and other volumes the composition Na_2SiO_3 in such proportions as to give the complete glass composition. If randomness were maintained, the x-ray diffraction technique would not be able to show this segregation.

Much may be learned about the coordination number of cations in the glassy structure from its color (see Chapter 17).

Network formers. It has been known for a long time that certain elements and compounds could be produced in the glassy state, but it was not until the classic paper of Zachariasen appeared in the year 1932 that the mechanism of glass formation was made clear. Dr. Zachariasen stated four rules that should be fulfilled before an oxide can be considered a glass former. These are: (1) each oxygen atom must not be linked to more than two cations; (2) the number of oxygen atoms around any one cation must be small; (3) the oxygen polyhedra must share corners, not edges, to form a three-dimensionsal network; (4) at least three corners of each must be shared.

On the basis of these rules, the following oxides should in themselves form glasses: SiO_2, B_2O_3, GeO_2, P_2O_5, and As_2O_5. Actually, all of these oxides will form glasses. In addition, As_2O_3, Sb_2O_3, and beryllium fluoride will form glass by the same criteria.

There are other materials with long-chain molecules that form glasses, such as the elements sulfur, selenium, and tellurium, and the compounds meta-phosphoric acid, zinc chloride, and some of the sulfides. Among them only As_2S_3 glass is important as an infrared-transmitting optical material.

A number of workers in the glass field have classified the cations into three classes: the network formers that can form a glass by themselves, the modifiers that cannot form glasses but can enter into the holes in the network, and the intermediates that at times may enter into the glass network to a limited extent. It has also been shown that the bond strength between a cation and an oxygen atom is an indication of glass-forming ability; the stronger bonds are the glass formers and the weaker ones the modifiers. Table 12.1, as worked out by Sun (1947), shows the single-bond strength of some cations as related to their glass-forming ability. It will

Table 12.1 Calculated bond strengths of oxide
constituents in glasses

	Cation	Valence	Coordination no.	Single bond strength
Glass network formers	B	3	3	119
	Si	4	4	106
	Ge	4	4	108
	Al	3	4	101–79
	B	3	4	89
	P	5	4	111–88
	V	5	4	112–90
	As	5	4	87–70
	Sb	5	4	85–68
	Zr	4	6	81
Inter- mediates	Zn	2	"2"	72
	Pb	2	"2"	73
	Al	3	6	53–67
	Zr	4	8	61
	Cd	2	"2"	60
Modifiers	Na	1	6	20
	K	1	9	13
	Ca	2	8	32
	Mg	2	6	37
	Zn	2	4	36
	Pb	4	6	39

be seen that the arrangement of the cations in groups according to bond strength gives a fair agreement with their glass-forming properties.

Network modifiers. The modifiers consist of the alkali ions, the alkaline earth ions, lead ions, zinc ions, and many others of less importance. As mentioned before, they weaken the network and loosen the bonding so that the glass has a softening point, is less chemically resistant, and has a greater coefficient of expansion.

Intermediate glass formers. There are other cations that may under suitable conditions enter the glass network itself as partial substitutes for the network former. In other cases, they will actually form the glass network with other nonglass-forming substances.

In the first case some cations, such as Al^{+++}, may partially substitute for Si^{++++} in the network. This is what happens in the clay minerals. In the second case, we have examples of such glasses as those found in the $MgO\text{-}CaO\text{-}Al_2O_3$, the $K_2O\text{-}CaO\text{-}Al_2O_3$, and the $BeO\text{-}Al_2O_3$ systems. None of the cations in these systems are glass formers in themselves. Al^{+++}, which in Table 12.1 is just on the borderline of the glass-former group, acts in these particular cases as a glass network former.

Methods of expressing glass composition. The student may well become confused by the many methods employed in the literature on the subject to express glass composition. Pincus brings this out clearly and his examples are assembled in Table 12.2 to give a direct comparison.

Table 12.2 Methods of expressing glass composition

Purpose	Expression	
Mol percent for glass	SiO_2	75.0%
	Na_2O	12.5%
	CaO	12.5%
Weight analysis	SiO_2	75.3%
	Na_2O	13.0%
	CaO	11.7%
Batch to yield 100 pts. of glass	Sand	75.4
	Soda ash	22.2
	Limestone	20.9
		118.5
Enamel frit	Sand	63.6%
	Soda ash	18.8%
	Limestone	17.6%
		100.0%
Empirical formula (glazes)	Na_2O 0.5 SiO_2 3.0	
	CaO 0.5	
Ionic formula	$Na_{0.33}$ $Ca_{0.17}$ Si $O_{2.3}$	
	or	
	$Na_{0.14}$ $Ca_{0.07}$ $Si_{0.43}$ O	
	$\underbrace{\qquad}$ $\underbrace{\qquad}$	
	m = 0.21 $\quad$ n = 0.43	
	glass network $\quad$ glass network	
	modifier $\qquad$ former	

Single-component glasses. The most important of the single oxide glasses is silica. This consists of a three-dimensional network of silicon-oxygen tetrahedrons bonded at every corner, so that each silicon is bonded to four oxygens and every oxygen is bonded to two silicons, giving an ionic expression of $Si_{0.50}O$. Because of the complete and strong bonding, silica glass has a high softening point, high viscosity, low coefficient of expansion, and chemical inertness.

Boric oxide glass, B_2O_3 or $B_{0.67}O$, has a weaker structure than SiO_2, since B^{+++} can surround itself with only three oxygens in triangular coordination. Thus the network approaches a plane, a type with weak forces in one direction. Hence boric oxide glass has a low softening point, is soluble in water, and has a high expansion coefficient. On the other hand, it is a very stable glass with regard to devitrification.

Phosphorous pentoxide glass, P_2O_5 or $P_{0.40}O$, forms tetrahedra with oxygen in the same way as silica, but since there are four O^{--} to one P^{5+}, only three unsaturated valence bonds are available. Thus one oxygen out of the four is linked to only one cation. This means that only three corners of the tetrahedrons are joined, which accounts for the low softening point and hygroscopic properties of this glass. Like B_2O_3 glass, it does not readily devitrify.

Binary glasses. Sodium silicate glasses are well known in the range $Na_2O\cdot3SiO_2$ ($Na_{0.29}Si_{0.43}O$) to $Na_2O\cdot SiO_2$ ($Na_{0.67}Si_{0.33}O$). As the silica decreases, the tendency to devitrify increases. The Na^+ satisfies the extra O^{--} bonds in the network, but the forces are weak and consequently these glasses are soluble and have low softening points.

The calcium or other alkaline earth silicate glasses are very interesting as they separate into two immiscible liquids. One of these is nearly pure silica and the other, lime-silica, is composed of $71SiO_2$, $29CaO$ ($Ca_{0.17}Si_{0.41}O$). Binary glasses of silica with either lead or zinc oxide occur over much wider ranges of composition than would be expected, and this is explained by the polarizing ability of Pb^{++} and Zn^{++}. P_2O_5 forms glasses readily with the alkaline earths or Al_2O_3.

Ternary glasses. The most important glass is, of course, soda-lime-silica, which forms glass over a considerable range. However, the useful glasses are confined to a small field about $Na_2O\text{-}CaO\text{-}6SiO_2$.

3. Elastic and Viscous Forces in the Glass Network

We have given in the preceding sections of this chapter something of the atomic arrangement in the glass structure. Now it will be instructive to examine glass in another way, that is, by a study of the integrated forces acting in the glass network.

Thermal expansion. Figure 12.1 shows a linear thermal expansion curve of a normal soda-lime glass. The curve represents an annealed specimen, which expands at a uniform rate until region *a* is reached, where the coefficient increases by a factor of

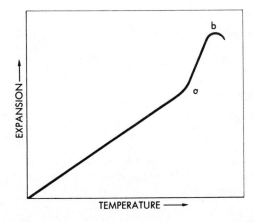

Fig. 12. 1 Expansion of a normal glass.

3 or 4. This zone of inflection is known as the transformation point. As the temperature range goes higher, point *b* is reached, which is known as the softening point. This expansion is completely reversible below point *a*.

There have been many theories propounded to account for this change in slope in the transformation region. This change is never sharp and varies with the rate of heating. It has been shown in the previous sections that glasses with weak bonds have higher expansion coefficients than stronger bonded networks. For example, B_2O_3 glass has thirty times the expansion coefficient of SiO_2 glass. It would seem reasonable to believe that, as the temperature is increased, thermal agitation weakens some of the bonds when an energy state corresponding to the temperature range *a* in Fig. 12.1 is reached and thus causes a gradual increase of the expansion coefficient.

Viscosity. The viscosity of glass is an important characteristic in working and annealing, and therefore its measurement is of practical as well as theoretical importance. The curve of Fig. 12.2 shows the viscosity of a normal soda-lime glass over a wide range of temperature. On the viscosity scale are noted several regions of importance to the glass maker:

Annealing range	$10^{12.5} - 10^{13.4}$ poises
Softening point	$10^{7.6}$
Working range	$10^4 - 10^{7.6}$
Melting range	$10^{1.5} - 10^{2.5}$

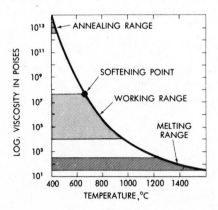

Fig. 12.2 Viscosity of a soda-lime-silica glass.

Extrapolating the viscosity curve to room temperature gives a value of 10^{27} poises. Undoubtedly stressed glass will flow at room temperature, but at a rate so slow that a lifetime would be needed to measure it.

Flow properties. When stress is applied to a glass heated to the annealing range, it will flow somewhat rapidly at first and then at a slower but steady rate (Fig. 12.3). If the stress is then released, the glass will at once contract and finally reach a constant length, which is greater than the original. If the straight line is extended to zero time, the intercept *a* will be the elastic deformation, and if the same is done when the stress is released, the elastic return *b* will be equal to *a*. It would then appear that the forces in the glass network are of two kinds; first, an elastic force caused by stretching of the stronger bonds, and second, a viscous force caused by a continuous breaking and reforming of the weaker bonds. In other words, there must be discrete groups of atoms that are only distorted elastically, and flow must take place between these groups. This flow may be duplicated by the simple model in Fig. 12.4, where the spring represents the elastic forces and the dash pot the viscous forces.

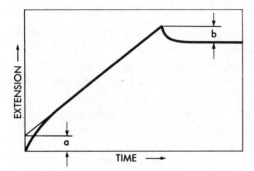

Fig. 12. 3   Dependence of flow rate on stress for a heated glass.

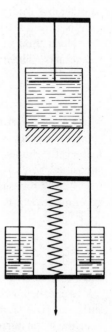

Fig. 12. 4   Mechanical model to show the flow in glass.

A glass at constant temperature may be stressed at several values and the uniform flow rate determined from the slope of the curves. If these rates are then plotted against stress on logarithmic paper, a straight line will result. The slope of this line will be n, for

$$\log\frac{dl}{dt} = \log k + n \log \sigma$$

$$\frac{dl}{dt} = k\sigma^n.$$

When $n = 1$ the flow is viscous, which is the case for glasses.

As shown in Fig. 12.3 the flow rate of glass at constant temperature and stress changes with time, and therefore the measured viscosity changes with time, eventually approaching an equilibrium value from either above or below the final stress value. Looking at it in another way, perhaps the true viscosity does not vary, but rather a combination of viscosity and a transient energy factor are measured.

Annealing. Annealing of glass is an important part of the manufacturing process. It is done to relieve the stresses that might be dangerously high in the case of air cooling. The heavier the glass object the slower must be the cooling rate. On the other hand, annealing does more than relieve the cooling stresses; it also permits the atomic structure to settle down into a more stable state as evidenced by higher density and higher index of refraction. In other words, in a quickly cooled glass considerable disorder is frozen in. Through annealing, the more compact arrangement is arrived at.

Much thought has been given to the mechanism of annealing. The well-known Adams and Williamson law states that

$$\frac{1}{\sigma} - \frac{1}{\sigma_0} = At,$$

where σ is the stress at time t, σ_0 is the initial stress at zero time, and A is a constant varying with temperature and composition. This equation fits the facts well at the upper end of the annealing range but is not satisfactory at higher viscosities.

Lillie proposes a relation which in many ways gives better correlation with experiment. This is

$$-\frac{d(\log \sigma)}{dt} = \frac{M}{\eta},$$

where σ is stress, t is time, M is the elastic modulus at the temperature of annealing, and η is viscosity at temperature of annealing.

Annealing will permit the reduction of stresses to any desired degree if sufficient time is taken. A bottle may be sufficiently annealed in a few hours, but a large telescope mirror takes months.

Quenched glass. It has been found desirable in some cases to cool a glass rapidly by an air blast to introduce high stresses in a predetermined manner. For example, sheet glass may be cooled on the faces, which brings them at once below the flow temperature, while the center of the sheet is still hot and deformable. The subsequent cooling of the center will then put the surfaces in compression, as shown by curve *a* in Fig. 12.5. Should a sheet annealed in the usual manner be stressed by bending, the forces across the thickness would be as shown in curve *b* and breaking would occur on the tension side, since glass is much stronger in compression than in tension. If the same load is now applied to the quenched glass, the internal stress distribution is obtained by the sum of the residual stress *a* and the applied stress *b*, which gives curve *c*. It will be seen that in this case the tension stress at the surface, which controls the breaking, is changed to compression and the total strength is greatly increased. Another factor in the use of chilled glass is the type of break; instead of the usual jagged fragments it breaks into small cubes.

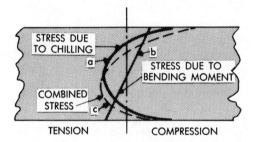

STRESS DUE TO CHILLING

a

STRESS DUE TO BENDING MOMENT

b

COMBINED STRESS

c

TENSION   COMPRESSION

Fig. 12.5   Stresses in a chilled glass sheet.

4. Devitrification

As far as we know, all inorganic glasses may be transformed either partially or completely into the crystalline state if they are held at temperatures in what is called the devitrification range. This range varies with the glass composition, but in general occurs at a temperature which gives the glass a viscosity of 10^6 poises. In some cases, the devitrification happens so rapidly that the glass is unable to go through the normal forming processes, whereas others are so stable that many hours of heating are needed to start crystallization.

Although devitrification is a troublesome factor in most glass production, it can be an advantage when carefully controlled. Examples of this kind are opal glass, mat glazes, and glass ceramics, products which are discussed later in this book.

The formation of crystals in the glass during devitrification occurs in two steps, as pointed out many years ago by Tamman. The first step is the formation of nuclei — small clumps of atoms arranged in the form of the crystal lattice; and second, the growth of these nuclei into crystals. The temperature zones for nucleus formation and crystal growth may overlap, in which case many sizes of crystals appear. On the other hand, when the zones are separate, there is the possibility of controlling the number and size of the crystals.

It is also found that a specific glass may devitrify into several species of crystal, either simultaneously or each in its own temperature range. Certain nucleating agents such as TiO_2 may direct and enhance the crystal development.

References

Zachariasen, W. H., The Atomic Arrangement in Glass, *J. Am. Chem. Soc.* **62**, 3841, 1932

Warren, B. E., X-ray Diffraction of Vitreous Silica, *Z. Krist.* **86**, 349, 1933

Morey, G. W., *The Properties of Glass*, Reinhold, New York, 1938

Kreidel, N., and W. A. Weyl, The Development of Low Melting Glasses on the Basis of Structural Considerations, *Glass Industry* **62**; I, 335; II, 384; III, 426; IV, 465; 1941

Sun, K., Fundamental Condition of Glass Formation, *J. Am. Ceramic Soc.* **30**, 277, 1947

Volf, M. B., *Technical Glasses*, Pitman, London 1961

Weyl, W. A., and C. E. Marboe, *The Constitution of Glasses*, 2 vols., Interscience, New York, 1962

Mackenzie, J. D., *Modern Aspects of the Vitreous State*, Vol. 3, Butterworths, London, 1964

Hair, M., The Structure of Alkali Silicate Glass, *J. Am. Ceramic Soc.* **52**, 677, 1969

Mozzi, R. L., and B. E. Warren, The Structure of Vitreous Silica, *J. Appl. Cryst.* **2**, 164, 1969

Tischen, R. E., Heat of Annealing in Simple Alkali Silicate Glasses, *J. Am. Ceramic Soc.* **52**, 499, 1969

Warren, B. E., *X-ray Diffraction*, Addison-Wesley, Reading, Mass., 1969

13

Glass melting

1. Introduction

The glass industry amounts to approximately one-third of the whole ceramic industry in value of annual production. The glass producers are progressive in research and development as well as in engineering, a fact that has done much to raise this industry to its present level.

2. Glass Compositions

Commercial glasses. The great bulk of the glass produced today is of the soda-lime-silica type with a small addition of alumina. This glass is used for containers and sheet glass. Increasing the alumina content adds to the chemical resistance and in some ways aids the working properties. Table 13.1 gives a typical analysis of a container glass. It is obvious that this composition may be produced from a large variety of raw material. Cost and availability would suggest silica sand, limestone, and soda ash, which are indeed the main ingredients, but it has been found desirable to add some of the soda in the form of the sulfate or nitrate, a procedure which is explained later in this chapter, and to add feldspar or nepheline syenite as a source of alumina. Although sheet glass is not decolorized, much container glass includes a decolorizer, as discussed later in this chapter.

Glass for tableware is quite similar to container glass, as shown in Table 13.1, except that it is lower in iron content and often contains barium oxide for greater brilliance.

The so-called flint glass or crystal is used for the highest grade of artware, especially when it is to be cut or engraved. This glass contains a considerable proportion of lead oxide (Table 13.1), which increases the index of refraction and thereby gives greater brilliance. The lead also makes the glass softer for cutting.

High-silica glasses have the advantages of low thermal expansion, high softening point, and good chemical resistance. Therefore they are much used for

Table 13.1 Glass compositions

Constituent	Window	Container	Pyrex	Textile fiber	Opal	Light flint	Heavy flint	Barium crown	Crown	Thermometer glass
SiO_2	72.1	72.5	80.5	54.0	65.8	67.4	46.1	59.1	72.2	67.5
B_2O_3			11.8	10.0				3.0	5.9	2.0
Al_2O_3	1.1	1.9	2.0	14.0	6.6	1.7	0.1	0.1		2.5
Fe_2O_3	0.2	0.1								
As_2O_3			0.7				0.4	0.3	0.2	
ZnO						3.9		5.0		7.0
CaO	10.2	9.8	0.3	17.5	10.1	0.4	0.1	0.1	2.1	7.0
MgO	2.6	0.1	0.1	4.5					0.1	
BaO		0.7						19.3		
PbO						10.7	45.1			
K_2O		0.8	0.2		9.6	0.1	6.8	9.7	13.9	
Na_2O	13.6	13.7	4.4		3.8	15.1	1.7	3.2	5.2	14.0
Sb_2O_3									0.1	
SO_3							0.1		0.1	
F_2					5.3					

laboratory and cooking ware. The general composition as shown in Table 13.1 indicates a low alkali content, the virtual elimination of alkaline earths, and the use of boric oxide to achieve a reasonable temperature for melting and working, although these temperatures are considerably higher than for container glass.

Fiber glass of the textile variety, because of its enormous surface area, must be particularly stable in respect to atmospheric moisture. Therefore a completely alkali-free glass is used for this purpose, as shown in Table 13.1. This glass has a rather high softening temperature, but also has viscosity properties that permit fibers to be drawn.

Optical glasses. There is not space here to discuss the thousands of glass compositions designed for optical use; however, a few typical compositions are shown in Table 13.1. The lens designer desires glasses not only with a considerable range of index of refraction and dispersion, but the greatest possible range in the ratio of the two. To illustrate this, the chart of Fig. 13.1 is given. Tremendous strides have been made in optical glasses during and since World War II, as shown by the added fields in the diagrams. The use of fluorides and rare elements has produced many excellent glasses. This work has been accomplished largely by the understanding of the principles of crystal chemistry, and great credit must go to the

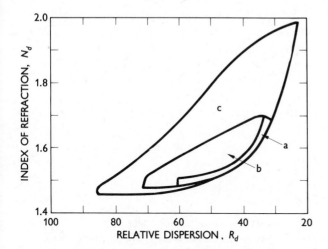

Fig. 13.1   Properties of optical glass: (a) before 1880; (b) from 1880 to 1934; (c) from 1934 to 1970. N_d is the index of refraction, R_d is the relative dispersion.

Eastman Kodak Company for much of this development. With these new glasses, it is possible to design simpler and more highly corrected photographic objectives than were previously possible.

Special glasses. There are many special glasses used, but only a few can be mentioned here. One of these is opal glass, used for containers and light diffusers. The production of opaque and translucent glasses will be discussed in Chapter 14. A commercial opal composition as shown in Table 13.1 contains fluorine, which produces crystals on cooling.

   The various colored glasses will be taken up in Chapter 17; in this field extensive production is confined principally to green and amber for beverage bottles.

3. Mechanism of Melting

The melting process is an interesting one, since it consists of many types of thermochemical process. In this discussion we will confine ourselves to a typical container glass. It should be remembered that the object of melting is to convert the batch materials into a homogeneous glass of constant properties for the forming operation, in the most economical manner.

Batch. A typical batch for container glass is shown in Table 13.2. It will be seen that besides the main ingredients of sand, limestone, and soda ash, there are several other materials. The feldspar is a source of alumina, while at the same time it adds alkalis and silica. The sodium nitrate, of course, adds soda to the glass, but it also acts as an oxidizing agent when it decomposes. The arsenic is a fining agent. The

Table 13.2
Typical container glass batch

Sand	1000 lb
Soda ash	342 lb
Limestone	262 lb
Feldspar	128 lb
Barytes	17 lb
Sodium nitrate	5 lb
Arsenic trioxide	0.5 lb
Manganese sulfate	0.160 oz
Cobalt oxide (Co_3O_4)	0.015 oz
Cullet	150 lb

manganese is a physical decolorizer that produces a pink color complementary to the green of ferrous iron. Iron is largely in the reduced state, even with an oxidizing agent, since the equilibrium goes in the direction of FeO at high temperatures.

Methods of batching. The large amounts of material to be handled in a glass plant require a carefully laid out system. The raw materials arrive in railroad cars or trucks and are stored in large silos. From these each batch is automatically weighed out, mixed, and placed in a series of containers moving on a conveyer to the glass tanks. The container system is used to minimize segregation in the batch which contains a large range of particle sizes.

Melting process. The melting process is an involved series of reactions that take place as the temperature rises and continue with time after the temperature has reached its maximum. It should be realized that most of these reactions occur simultaneously, but as yet we have no reliable data on the exact life history from the raw batch to the refined glass.

Fining glass. This operation consists of dissolving the last grains of silica and absorbing the fine bubbles of gas present in the melt — a slow process. The early glass-makers assisted the fining by throwing pieces of raw potato into the melt to form large bubbles of steam which, rising, tended to carry the smaller bubbles with them. It was found later that materials like antimony or arsenic trioxide added in amounts of 0.05% of the batch greatly speeded up the time taken for the bubbles to disappear.

The action of fining agents is still not fully understood, but active research is bringing many facts to light. Workers at Sheffield University have found it possible to analyze the gas contained in bubbles as small as 10 microns in diameter by using a chromatograph.

It has been shown that the bubble initially contains CO_2 from the decomposition of the carbonates, but as time goes on the oxidizing agents (e.g., the nitrates) supply O_2 and this gradually replaces the CO_2 inside the bubble. Oxygen is highly diffusible in the melt, so the bubble rapidly decreases in size and finally disappears if there is somewhere for the oxygen to go. This is where the fining agent comes in: The reaction $R_2O_3 \rightarrow R_2O_5 - O_2$ provides, as it were, an oxygen sink. There is a great incentive to find ways to eliminate the expensive and often troublesome fining agents, but up to the present no alternatives have been found.

Stones in the glass. The glass maker is sometimes troubled by glass stones, small undissolved particles that go all the way through the melting process. These stones can be identified by the petrographic microscope of x-ray-diffraction methods. They usually consist of silica from the batch, particles of refractory from the tank blocks, drips from the crown, or local devitrification. After the source has been found, steps can be taken to eliminate them.

Decolorizing glass. In normal container glass the color comes from the iron oxide in the batch, which runs from 0.02 to 0.04%. As shown in Fig. 13.2, both the ferric and ferrous ions absorb in the visible spectrum, the former giving a yellow color and the latter a green-blue color. As the absorption of the ferric ion is only 10% of that of the ferrous ion it is advantageous to keep the glass melt in the oxidized condition. Glass without decolorizer has about 70% of its iron in the ferrous state,

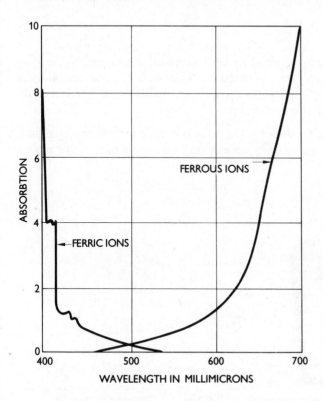

Fig. 13. 2 Absorption of ferric and ferrous ions in glass.

but an oxidizing agent such as cerium trioxide will lower this to about 6%, according to the equation

$$2 \, CeO_2 + 2 \, FeO = Ce_2O_3 + Fe_2O_3.$$

The remaining yellow color may then be masked by a complementary purple, supplied by manganese, neodymium, or selenium oxides. A container batch with 0.03% iron oxide would have added to it 0.007% CeO_2 and 0.0007% Nd_2O_3, or the latter could be replaced with 0.0016% manganese sulfate or 0.00003% cobalt oxide. Decolorizing is a troublesome operation requiring careful temperature and atmosphere control.

Stirring glass. Most optical glass, which must be extremely homogeneous, is stirred to break up stria during the firing and first part of the cooling period. In pots the stirrer is a round, water-cooled tube covered with refractory, which extends down into the glass and is moved around in a pattern that covers all the glass volume. When the temperature falls to 1100°C for crown or 950° for flint the stirrer is removed. In the case of continuous tanks the stirrers, if used, are in the forehearth just ahead of the feeder.

4. Melting Equipment

Introduction. Before the middle of the eighteenth century glass was melted in pots made of special clays. As the demand for glass increased, rectangular reverberatory furnaces called day tanks were introduced. Later, when automatic forming machines came into use, still greater capacity was needed and this led to the development of the continuous melter.

Pots. A typical covered pot is shown in Fig. 13.3. It is made of "pot clay" and grog, giving a body with high enough hot strength to hold the pressure of the molten

HOLE FOR WORKING OUT
GLASS AND CHARGING

FRONT
WALL

Fig. 13. 3   Cross section of a glass pot.

glass. Pots have been and to some extent still are hand-molded, a slow and exacting process, but now most glass pots are slip cast.

Pots are still used for optical glass, colored glass, and art glass. They range in capacity from 500 lb of glass up to 4000 lb, and the time from filling to working may be as little as 24 hours, at temperatures of around 1400°C. The life of the pot is several months with periodic charging and working out. As the whole furnace cannot be cooled down to change one pot, it is necessary to preheat the new pot, break down a portion of the furnace wall, remove the white-hot old pot, and replace it with the new preheated one — a hot and exacting operation.

Some optical glass is now melted in platinum-lined pots. Although the platinum is very costly, it may be reworked with a loss of less than 10% of its value. The glass produced in platinum pots is especially homogeneous, since there is no solution of refractory as in the case of clay pots.

Day tanks. Obviously, the cost of melting in pots is high, so a less expensive method for larger quantities was developed in the form of the day tank. This melts glass in batches like the pot, but there is a fundamental difference. Here the glass is heated entirely from the free surface, and the refractory temperature is always lower than the mean glass temperature. This allows less reaction between the glass and refractory but causes large temperature gradients in the glass and tends to produce inhomogeneity. Glasses more transparent to infrared radiations show more even temperatures, as might be expected.

Container-glass tanks. These tanks hold 200 to 300 tons of glass and can operate for an average of 4½ years without a shut-down, melting 0.3 tons of glass per square foot of melting area per 24 hours. Figure 13.4 shows such a tank with the important features indicated. The batch is fed into the tank by a continuous conveyer at one or more openings called dog houses and spreads out over the molten bath in a layer 5 or 6 inches thick. The heat is supplied by side burners with exhausts on the opposite side. The air is heated in checker chambers below each burner port and the direction of flow reversed about every 20 minutes. The fuel is natural gas or oil.

The glass is hotter and more fluid in the center of the tank and quite viscous along the sides and bottom where it is cooled by the refractories. At the end of the melting chamber the glass passes through a throat orifice near the bottom of the tank, which prevents floating material from passing into the fining area. The throat is at the bottom of a bridge wall, often water cooled, that crosses the tank and supports a checker-type shadow wall; this arrangement allows just enough radiant heat to enter the fining zone. The molten glass cools in the fining chamber until it reaches the right viscosity to enter the forehearth (or forehearths) at the end of the tank.

The side walls, end walls, and the bottom of the tank are made of large refractory blocks, which may be of fire clay and grog but are usually fusion-cast blocks of alumina or a zirconia-alumina-silica mixture. The bottom is sometimes

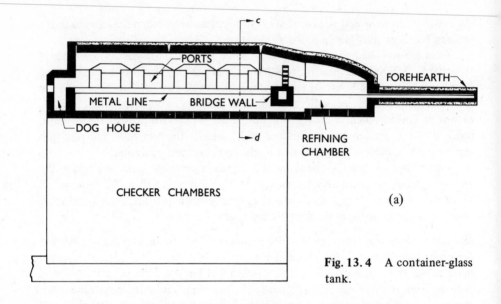

Fig. 13. 4   A container-glass tank.

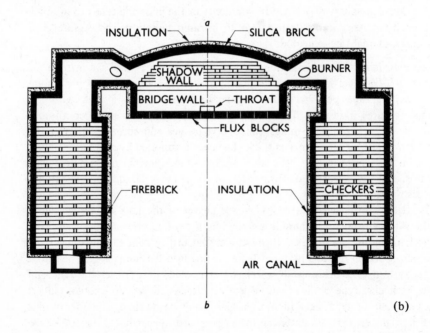

paved with dense zircon tiles and the roof, 9 to 13½ inches thick, is silica. Recent tanks have been completely insulated, but coolers are used along the metal line (top of the glass melt) in the melting section. Modern container tanks have melting areas of 1000 to 1500 sq. ft and the glass depth is 3 to 4 feet.

In Europe some container glass has been melted in electric tanks where the heat is generated in the molten glass by resistance heating. Electrodes of graphite or molybdenum are used. Since natural gas has become available in Europe, electric melting has not looked so attractive, although electric boosters are often used in fuel-fired tanks to give increased capacity. Bubblers to blow air into the bottom of the melt are used to some extent for circulating the glass, but there is a difference of opinion as to their usefulness.

Sheet-glass tanks. These tanks are much larger than the container tanks, often holding 1500 tons of glass. In Fig. 13.5 is shown a typical tank, very much like the container tank except that it is longer and in place of the bridge wall and throat it has a floater to hold back scum. In place of the shadow wall an adjustable curtain wall controls the temperature in the fining zone. One of these tanks will supply 200 to 300 tons of glass in 24 hours.

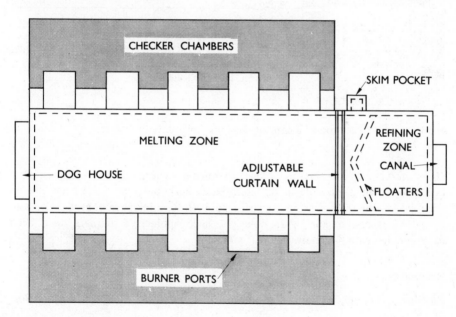

Fig. 13.5 Plan of a sheet-glass tank.

Both the container- and sheet-glass tanks have recuperation to save fuel but even so have a low thermal efficiency of only 15 to 25%. The greater loss through the fused tank blocks in the modern tank is about balanced by the added insulation.

Tanks for optical and ophthalmic glass. These tanks are small, turning out 100 to 400 pounds per hour. The melting, as shown in Fig. 13.6, is carried out in a gas-heated melting chamber, usually made of fusion-cast blocks in walls, hearth, and crown. In some cases molybdenum or tin oxide electrodes are used in the melting zone to give increased capacity. The glass passes from the melting chamber into a platinum cylinder refiner heated by silicon carbide heater rods. From the refiner the glass flows into a platinum stir chamber, where it is homogenized, and then passes down a tube to be sheared into gobs. Such a tank may have a third of a million dollars worth of platinum in it and generally will run two years without rebuilding.

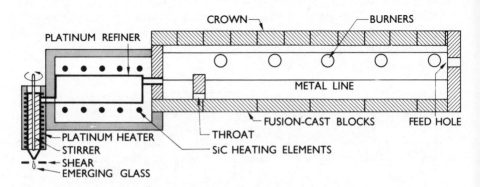

Fig. 13. 6   Cross section of a continuous tank for ophthalmic glass.

Other methods of glass melting. Many attempts have been made to melt glass more efficiently than is possible in the continuous tank, but none have been able to supplant it. One method of melting slag and rock for mineral wool production uses a water-jacketed cupola with coke as fuel. Submerged burners have been experimented with, but show no great promise.

References

Riebling, E. F., Silicon Dioxide Dissolution in Molten Glass, *Bull. Am. Ceramic Soc.* **48**, 766, 1969

Green, C. H., and D. R. Platts, Behavior of Bubbles of Oxygen and Sulphur Dioxide in Soda-lime Glass, *J. Am. Ceramic Soc.* **52**, 106, 1969

Helzel, M., Gas Chromatography Analysis of Gaseous Inclusions in Glass, *Bull. Am. Ceramic Soc.* **48**, 287, 1969

Onada, G. Y., Jr., and S. D. Brown, Low Silica Glasses Based on Calcium Aluminates, *J. Am. Ceramic Soc.* **53**, 311, 1970

Cable, M., and M. A. Harvon, The Action of Arsenic as a Refining Agent, *Glass Tech.* **11**, 49, 70

Stanforth, J. E., Oxide Glass Formation from the Melt, *J. Am. Ceramic Soc.* **54**, 61, 1971

14

Glass forming

1. Introduction

Early glass vessels were made by covering a sandstone core with fibers or rods of glass, which were then worked together. Later the blowpipe greatly aided the glass-maker and even now art glass and some chemical ware is produced in this way.

Since the start of this century there has been rapid progress from the single-mold, hand-operated press to rotary table presses and then to the straight-line machines. Anyone observing a modern glass plant operating cannot help but be impressed with the rate and smoothness of the operation. Of course, behind this lies a tremendous amount of both ingenious and sound engineering.

2. Forming Methods

Feeders. The first step in automatic glass forming is to supply a glass that is uniform in all respects. This is the function of the forehearth, which takes glass from the fining chamber, brings it to the correct temperature, and stirs it. Figure 14.1 is a diagram of a typical forehearth, with multiple burners, several revolving stirrers, and three gob feeders. Details of the feeder are also shown, in which the height of the

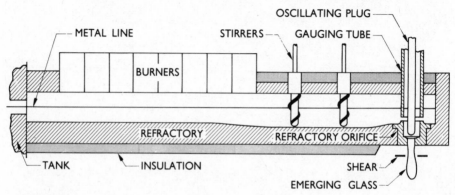

Fig. 14.1   Cross section of a forehearth.

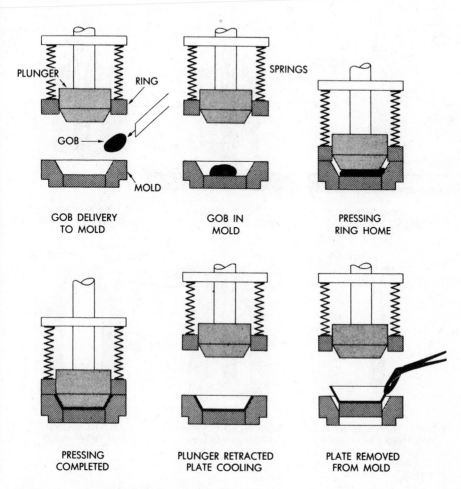

Fig. 14. 2   Steps in pressing a glass dish.

tube regulates the rate of flow and the oscillating plunger attenuates the stream to allow more efficient shearing at its narrowest point. The feeder operates at a rate of 15 to 75 gobs a minute. A shear mark is left on each gob, so means are provided to orient the gob as it slides down to the mold. A feeder can handle 20 tons of glass in 24 hours.

Pressing. This method is used for plates, tapered tumblers, and other thick-walled objects. It is believed that the first glass press was developed by a carpenter at the Sandwich Glass Company in the year 1816. This was the forerunner of the automatic glass-forming machine which later revolutionized the industry.

Figure 14.2 shows the steps in pressing a dish. The molds are made of a special cast iron surfaced on the inside with a firm, smooth layer of carbon developed from oily coatings. A typical coating medium is made up as follows: linseed oil, 100 g;

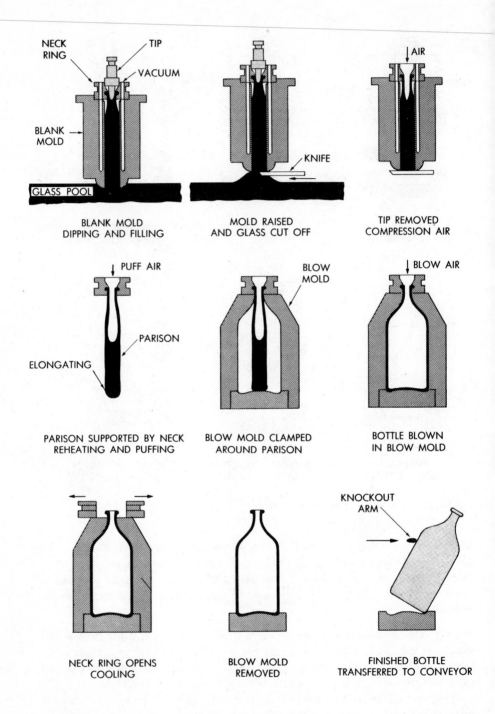

Fig. 14. 3   Vacuum and blowing process (Owens Machine).

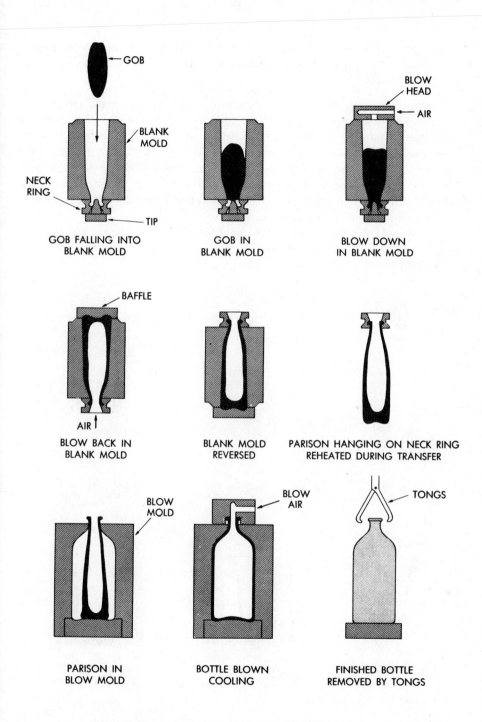

Fig. 14. 4   Blowing and blowing process (Lynch Machine).

rosin, 200 g; and fine cork dust, 100 g. The mold must be kept at the correct temperature, for if it is too hot, the glass will stick and if too cold, it will not form an even surface.

Blowing. Blowing is used for most containers, such as bottles and jars. Several processes are used, but the ones illustrated in Figs. 14.3 to 14.5 clearly show the principles of forming. It will be seen that there are two steps in each method, the first a forming of the parison or temporary shape from the gob, and the second the blowing of the parison to fit the inside of the blow mold. While the first step varies with the different methods, the second is the same in all of them. During the blowing, the thicker portions of glass retain their heat longer than the thin portions and therefore flow more, thus producing an even wall thickness.

The molds here are special cast iron maintained at a temperature of about 400°F and coated with a mold paste. The machines that operate the molds are fully automatic, but they will not be described here.

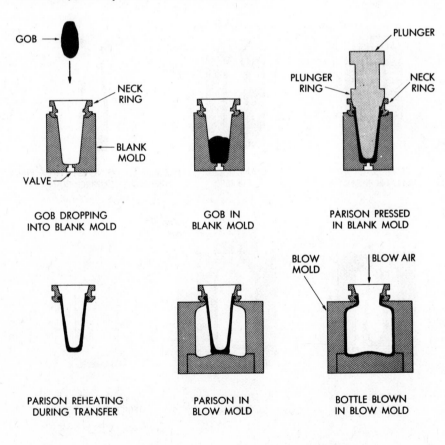

Fig. 14. 5   Pressing and blowing process (Lynch Miller Machines).

Drawing. Sheet glass is now formed in this country by drawing a continuous sheet from a pool in the feeder of the glass tank, as shown in Fig. 14.6. The maintenance of exact temperatures in the sheet at each point in the cycle is very important. Some sheet is rolled as shown in Fig. 14.7, the glass running over a refractory weir. Tubing or rods may also be drawn, either intermittently or continuously.

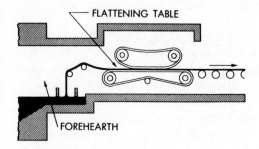

Fig. 14. 6   Drawing sheet glass.

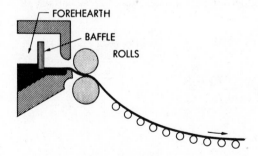

Fig. 14. 7   Rolling sheet glass.

Plate glass produced as in Fig. 14.7 is further treated by an ingenious process invented by Pilkinton Bros., which produces sheets with a flatness and surface quality approaching that of expensive ground and polished plate glass. Figure 14.8 shows the general scheme of operation. The sheet from the tank goes through a gas seal and floats along on a bath of molten tin with radiant heaters above and

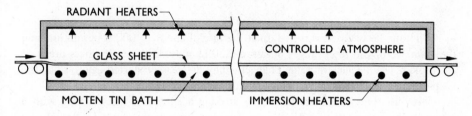

Fig. 14. 8   Float glass process.

immersion heaters below so the glass settles down into an even ribbon. It is then allowed to cool gradually, passed through another gas seal, and into the *lehr* for annealing. A reducing atmosphere must be retained in the float chamber to prevent oxidation of the tin. At present all automobile glass is produced in this way.

Fiber forming. The use of glass in the form of fibers has been very extensive. This material may be divided into two types: (1) the continuous filament for textile use and (2) the discontinuous fiber employed in heat insulation, filters, and plastic reenforcement.

The continuous filaments are drawn from a series, often 204, of platinum spinnerets set in the bottom of a platinum heating chamber called a bushing. The glass is fed into this chamber in the form of glass marbles at a rate that keeps the glass level constant. A cross section of a spinneret and fiber is shown in Fig. 14.9. The fiber is pulled from the bottom of a drop of glass held in the mouth of the spinneret and the flow of glass through the passage of about 0.07-in. diameter simply replenishes the drop. As the viscosity of the glass must be close to the correct value, the temperature control must be exact. The temperature is maintained by electric resistance heating of the platinum chamber, although one manufacturer uses high-frequency induction heating that has the advantage of greater uniformity.

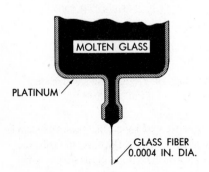

MOLTEN GLASS

PLATINUM

GLASS FIBER
0.0004 IN. DIA.

Fig. 14.9   Cross section of a bushing used for forming glass-textile fibers.

The bundle of fibers is reeled up at the rate of about 15,000 ft per minute and the diameter of the fiber ranges between 0.0002 and 0.0005 in. If fibers break or are broken when reels are being changed, they are readily started again, since the drop on the spinneret falls by gravity and draws a new fiber behind it.

Discontinuous fibers are blown by striking a stream of glass or slag with a high-velocity jet of steam. All blown fiber contains some shot, 15 to 25% by weight, and many believe that these shot are formed first and are then drawn out into fibers as they move through the air. This is not always so; the fibers are often formed first, and if they are not chilled rapidly enough, surface tension draws them up into spheres. Glass fibers are blown from a multiple stream of glass running from a series of nozzles in the bottom of an electrically heated platinum melting chamber. In this case the air or steam blast is parallel to the glass stream and accelerates it to a very high velocity.

Some fibers are produced by having the glass or melted rock stream fall on a rapidly revolving disk; this produces a low shot content.

3. Finishing and Annealing

Fire polishing. Laboratory ware or tableware, after forming, is usually smoothed on the seams and edges by applying localized heat. This lowers the viscosity to the point where surface tension forces can level the surface.

Grinding. Light tableware and artware is often finished by grinding and polishing the edges on a wheel.

Annealing. After finishing, the glassware is cooled at a given rate in a *lehr* (annealing oven). Usually this is a continuous process with the ware passing through the long controlled-heat chamber on a moving mesh belt. Sheet glass is handled in the same way with the sheet moving at the drawing speed through a long lehr.

4. Grinding and Polishing

Grinding plate glass. This operation takes place on the face of large plates that are bedded down in plaster. Wet sand is used as an abrasive under revolving cast-iron laps. At first the sand is quite coarse, but as the grinding proceeds it becomes finer and finer until a very smooth, mat surface is produced on the glass.

Polishing. The finely ground plate-glass sheet is next polished by passing it under felt laps fed with a suspension of fine rouge (Fe_2O_3) in water. The mechanism of polishing glass is not completely understood, but it is generally believed that a thin layer of the surface actually flows owing to the high temperatures and pressures. Thus the humps are slid into the hollows to produce a level surface. Some think the glass is actually melted and that only refractory particles are capable of polishing. However, there is no known explanation of why rouge and cerium oxide are among the very few efficient polishing agents.

Optical glass is not only polished with felt laps but also with hard ones made from a special pitch. Care must be taken to eliminate all oversized particles in the suspension to prevent scratching.

References

Glass Glossary, *Bull. Am. Ceramic Soc.* **27**, 353, 1948

Davis, P., *The Development of the American Glass Industry,* Harvard University Press, Cambridge, 1949

Rhodes, J. R., and B. W. King, Jr., Leveling of Vitreous Surfaces, *J. Am. Ceramic Soc.* **53**, 134, 1970

Cassidy, D. C., and N. A. Gjostein, Capillarity-induced Smoothing of Glass Surfaces by Viscous Flow, *J. Am. Ceramic Soc.* **53**, 161, 1970

15

Glazes

1. Introduction

A glaze may be defined as a continuous adherent layer of glass, or glass and crystals, on the surface of a ceramic body. The glaze is usually applied as a suspension of the glaze-forming ingredients in water, which dries on the surface of the piece in a layer. On firing, the ingredients react and melt to form a thin layer of glass. The glaze may be fired at the same time as the body or in a second firing.

The main purpose of the glaze is to provide a surface that is hard, nonabsorbent, and easily cleaned. At the same time the glaze permits the attainment of a greater variety of surface colors and textures than would be possible with the body alone.

2. Methods of Expressing Glaze Compositions

Batch formula. The glaze may be specified by giving the weights of the ingredients. For example, a typical raw lead glaze is:

White lead	154.8 g
Whiting	30.0
Feldspar	55.7
Kaolin	25.8
Flint	48.0
Total	314.3

This list of ingredient weights is excellent for the person mixing the glaze, but it does not permit the technologist to visualize the nature of the glass structure formed, or to compare the glaze with another.

Equivalent formula. The glaze given in the previous section may also be expressed in molecular equivalents of the oxides, as first suggested to the ceramists by Seger, which gives:

$$
\begin{array}{lll}
0.6\ PbO & & \\
0.3\ CaO & 0.2\ Al_2O_3 & 1.6\ SiO_2 \\
0.1\ K_2O & &
\end{array}
$$

The RO or basic oxides (where R is any cation) are placed in the first column with a total of unity, the R_2O_3 or amphoteric oxides in the second column, and the RO_2 or acid oxides in the last column. Although this method of expressing glazes was developed long ago, it can be seen in the light of our modern knowledge of crystal chemistry that in general the constituents have been grouped respectively into glass network modifiers, intermediates, and formers.

Ionic formula. As shown in Chapter 12, a glass or glaze may be expressed in the ionic form with the network formers as unity or the total anions as unity. Thus our raw lead glaze becomes:

$$
Pb_{0.6}^{++}\ Ca_{0.3}^{++}\ K_{0.2}^{+}\ Al_{0.4}^{++}\ Si_{1.6}^{++++}\ O_{4.8}^{--},
$$

or

$$
Pb_{0.38}^{++}\ Ca_{0.19}^{++}\ K_{0.13}^{+}\ Al_{0.25}^{+++}\ Si^{++++}\ O_{3.0}^{--}.
$$

The silicon-oxygen ratio then is 0.33. This is of the proper magnitude for a stable glass, for if the anions are unity, we get

$$
Pb_{0.13}^{++}\ Ca_{0.06}^{++}\ K_{0.04}^{+}\ Al_{0.08}^{+++}\ Si_{0.33}^{++++}\ O^{--}.
$$

Then $m = 0.31$ and $n = 0.33$.

Classification of glazes as to composition. The following is a list of the common types of glaze:

1. Raw glazes (containing insoluble raw materials)
    a) lead-containing glazes
    b) zinc-containing glazes (Bristol)
    c) porcelain glazes

2. Fritted glazes (containing some glass before firing)
    a) lead-containing glazes
    b) leadless glazes

3. Vapor glazes (deposited from the vapor phase)
    a) salt glaze
    b) smear glaze

The composition ranges of these glazes conform in a general way to the chart of Fig. 15.1, redrawn from one by Holscher and Watts.

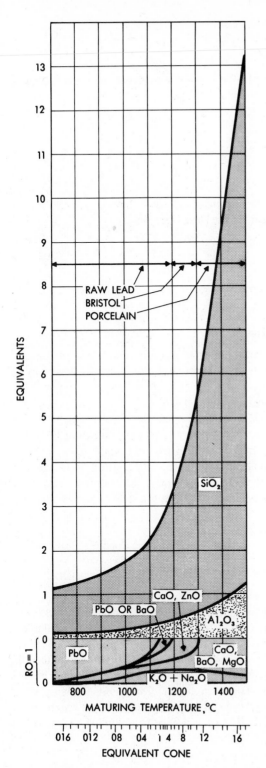

Fig. 15.1 Graphic representation of the composition of raw glazes. For example, for a raw lead glaze maturing at $1100°C$ it is: $0.3\,PbO$, $0.25\,K_2O$, $0.45\,CaO$, $0.3\,Al_2O_3$, and $2.0\,SiO_2$.

Classification of glazes as to surface. It is also possible to classify glazes according to surface characteristics, thus:

Glossy
Semimat
Mat
Surface crystalline
Vellum

Classification of glazes as to optical properties. Also they may be classified according to the nature of the interior of the glaze layer, as:

Transparent (clear)
Opaque (majolica, enamel)
Fine interior crystals (aventurine)
Large interior crystals (crystalline)

Of course, glazes may be readily classified as to color.

3. Methods of Compounding Glazes

Raw glazes. In ceramic literature, glazes are often given in the equivalent form, so that it is necessary to change this to the batch formula before the glaze can be made up. Even without a knowledge of chemistry anyone will find this quite simple, since it is really nothing but an arithmetical operation.

First, the term *equivalent weight* should be made clear. For example, potash feldspar, $K_2O \cdot Al_2O_3 \cdot 6SiO_2$ has a formula weight of 556.8 g as shown in Table A.2 in the Appendix. Thus 556.8 g of potash feldspar will yield one molecular equivalent (94.2 g) of K_2O, one molecular equivalent (101.9 g) of Al_2O_3, but six molecular equivalents (360.6 g) of SiO_2. Therefore, if it is desired to add one equivalent of SiO_2 by means of feldspar, not 556.8 g but rather 556.8/6 or 92.8 g of it would be added.

As an example, take a simple raw lead glaze such as:

$$0.1\ K_2O$$
$$0.8\ PbO \qquad 0.2\ Al_2O_3 \qquad 3.3\ SiO_2$$
$$0.1\ CaO$$

In converting to a batch formula some judgment must be exercised in choosing the raw materials. One could derive K_2O from K_2CO_3, but as this is soluble it would not be suitable for a raw glaze. Thus feldspar is the logical material. In the same way CaO could be obtained from several chemicals, but the carbonate is the most convenient. The PbO could come from red lead or white lead, but the latter is generally preferred as it stays in suspension better. The kaolin is a suspending medium, but not more than 0.15 equivalent should be added in the raw state or too much shrinkage will occur in drying. If more is needed to supply Al_2O_3, a portion of the kaolin should be calcined.

The student should get in the habit of carrying out the glaze calculation in a systematic manner, as shown in Table 15.1. The equivalents of each ingredient are multiplied by their respective equivalent weights to give the batch weights or percentage composition, as shown in Table 15.2.

Table 15.1

Original form	0.3 Na_2O	0.4 CaO	0.3 BaO	0.1 Al_2O_3	2.6 SiO_2
0.60 Frit	0.3 Na_2O	0.2 CaO	0.1 BaO	—	1.0 SiO_2
Remainder	0	0.2 CaO	0.2 BaO	0.1 Al_2O_3	1.6 SiO_2
0.20 Whiting		0.2 CaO	—	—	—
Remainder		0	0.2 BaO	0.1 Al_2O_3	1.6 SiO_2
0.20 Barium carb.			0.2 BaO	—	—
Remainder			0	0.1 Al_2O_3	1.6 SiO_2
0.10 Kaolin				0.1 Al_2O_3	0.2 SiO_2
Remainder				0	1.4 SiO_2
1.40 Silica					1.4 SiO_2
Remainder					0

Table 15.2

Equivalent	Ingredient	Approximate formula	Equivalent weight	Batch weight	Percent
0.1	Potash feldspar	$K_2O \cdot Al_2O_3 \cdot 6SiO_2$	556.8	55.7	12.9
0.1	Whiting	$CaCO_3$	100.1	10.0	2.3
0.8	White lead	$2PbCO_3 \cdot Pb(OH)_2$	258.5	206.5	48.1
0.1	Kaolin	$Al_2O_3 \cdot 2SiO_2 \cdot 2H_2O$	258.1	25.8	6.0
2.2	Flint	SiO_2	60.1	132.2	30.7
Total				430.2	100.0

The conversion of the batch formula back to the empirical formula is carried out by reversing the process, but in this case no judgment is required in selecting ingredients. For any exact calculation, the actual composition of the materials would have to be considered. For example, kaolin often contains some quartz and feldspar, and feldspar contains some quartz.

Fritted glazes. The calculation of fritted glazes is more complicated than that of raw glazes. The main purpose of fritting is to be able to use water-soluble materials by melting them together with other materials to form a relatively insoluble glass.

For example, boron compounds are nearly all soluble and must be made into an insoluble borate glass before being used in the glaze suspension. Secondary purposes of fritting are to obtain better working properties of the wet glaze, to distribute color more evenly, or to handle the lead in a less poisonous form.

In compounding a frit there are certain rules usually followed in selecting the composition so that the melted frit will form a glass, fluid enough to flow and at the same time sufficiently insoluble to be ground in water.

These rules are:

1. The ratio of the basic oxides to the acid oxides should be between 1 : 1 and 1 : 3. This means that the composition must be in the glass-forming range. If the B_2O_3 content is high, the ratio may be much larger

2. All soluble alkalis and boric oxide should be in the frit

3. The ratio of the alkaline oxides to the other basic oxides should not be much more than one

4. The ratio of B_2O_3 to SiO_2 should be less than one-half

5. The alumina content should be less than 0.4 equivalent

As an example, take the following glaze:

$$0.3\ Na_2O$$
$$0.4\ CaO \qquad 0.1\ Al_2O_3 \qquad 2.6\ SiO_2$$
$$0.3\ BaO$$

It is evident that the high Na_2O content will make it impossible to use feldspar without increasing the Al_2O_3 content too much. Therefore, most or all of the soda must come from soda ash (Na_2CO_3). Following the rules previously given, we find the following frit is reasonable:

$$0.30\ Na_2O \qquad\qquad\qquad 0.50\ Na_2\ O$$
$$0.20\ CaO \quad 1.0\ SiO_2 \qquad or \qquad 0.33\ CaO \quad 1.67\ SiO$$
$$0.10\ BaO \qquad\qquad\qquad 0.17\ BaO$$

The complete glaze is worked out as before (Table 15.3). The batch weights of the frit may be found as shown in Table 15.4. It should be kept in mind that the total weight given above as 310.1 is the weight of the raw materials in the batch, and not the resultant weight of frit. Therefore, not 0.6 × 310.1 g of frit are added to the glaze but rather 0.6 × 175.9, calculated on the basis of the oxides alone. Now the complete batch formula can be made up as in Table 15.5.

The frit may be melted in crucibles for small batches, but production lots are made in rotary frit furnaces or small glass tanks. It is quenched in water and ground to about 35 mesh. It can then be mixed with the remainder of the glaze batch and wet-milled for the correct length of time. One thing the glaze formula does not indicate is the fineness of grinding. The working properties of the glaze, as well as the maturing temperature, depend to a considerable extent on the fineness, so the milling conditions must be closely controlled.

Table 15.3

Original glaze	0.1 K$_2$O	0.1 CaO	0.8 PbO	0.2 Al$_2$O$_3$	3.0 SiO$_2$
0.1 Potash feldspar	0.1 K$_2$O	—	—	0.1 Al$_2$O$_3$	0.6 SiO$_2$
Remainder	0	0.1 CaO	0.8 PbO	0.1 Al$_2$O$_3$	2.4 SiO$_2$
0.1 Whiting		0.1 CaO	—	—	—
Remainder		0	0.8 PbO	0.1 Al$_2$O$_3$	2.4 SiO$_2$
0.8 White lead			0.8 PbO	—	—
Remainder			0	0.1 Al$_2$O$_3$	2.4 SiO$_2$
0.1 Kaolin				0.1 Al$_2$O$_3$	0.2 SiO$_2$
Remainder				0	2.2 SiO$_2$
2.2 Flint					2.2 SiO$_2$
Remainder					0

Table 15.4

Equivalent	Material	Equivalent weight	Weight	Percent
0.50	Sodium carb. (cryst.)	286.2	143.1	46.2
0.33	Whiting	100.1	33.0	10.6
0.17	Barium carb.	197.4	33.5	10.8
1.67	Flint	60.1	100.5	32.6
Total			310.1	100.0

Table 15.5

Equivalent	Material	Equivalent weight	Weight	Percent
0.60	Frit	175.9	105.5	38.4
0.20	Whiting	100.1	20.0	7.3
0.20	Barium carb.	197.4	39.5	14.4
0.10	Kaolin	258.1	25.8	9.4
1.40	Flint	60.1	84.0	30.5
Total			274.8	100.0

4. Application of Glazes

Properties of the glaze suspension. Anyone who has worked with laboratory-made glazes and compared them with a carefully developed commercial glaze will have seen at once the superiority of the latter in working properties. This is due to controlled particle size and correct selection of suspending clays.

The glaze slip should have these properties:

1. Low rate of settling
2. A high mobility (low viscosity) so that it will flow out into a smooth surface
3. A high yield point, not in the slip form but after it has lost a little water, so that the glaze layer will not slough off
4. A low drying shrinkage
5. A high elasticity in the dry condition
6. Little change of slip properties with aging

These conditions are met only through very careful control of specific gravity, particle size, pH of the suspension, and type of clay, and perhaps with added organic matter in the form of water-soluble gums.

Methods of application. Hand dipping of bisque ware into a tub of glaze slip was the only method used until World War I and even now some ware is glazed in this way, but it requires considerable skill to get an even glaze coating. Some ware such as small high-tension insulators are glazed in a dipping machine.

As pottery making became more industrialized the application of glaze by spraying came into use. The glaze slip is readily atomized in a spray gun with compressed air. As the tiny droplets strike the ware surface they flatten out into thin discs and quickly lose part of their water by diffusion and evaporation. Successive droplets build up, one upon another, to coalesce into a uniform layer. If the application rate is too high, the layer will run down vertical surfaces; if too low, a porous layer will be built up because the droplets do not join. Hand spraying complicated pieces, such as sanitary ware, requires great skill to give an even coating.

Automatic spraying on a conveyor line is now common practice. The ware may go through the spray booth on a belt, be turned over, and go through another spray booth. In other cases the ware, such as plates, is held on a three-pointed support and top and bottom are glazed at the same time.

In Europe for some time tiles have been glazed by passing them on a belt under a falling sheet of glaze slip like a waterfall (Fig. 15.2). As tiles in the United States are largely single-fired this process has been little used here, but it has recently been found possible to glaze green tile with the waterfall method so that this process will be more used.

An adaptation of the waterfall method is now used for tableware which supplies a veil or sheet of glaze slip at low velocity from slitlike nozzles directed on the revolving ware.

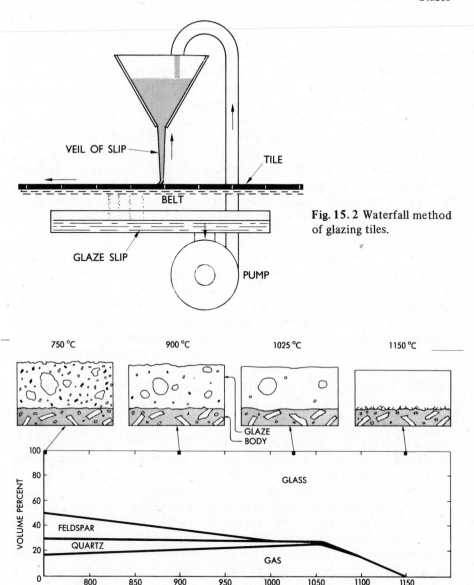

Fig. 15. 2 Waterfall method of glazing tiles.

Fig. 15.3 Life history of a fritted glaze. Below, the volumes of the various constituents; above, thin sections of the glaze and body at the various temperatures.

5. Firing the Glaze

Life history of a glaze. The diagrams in Fig. 15.3 show the stages of reaction in a simple fritted whiteware glaze — a process which is quite similar to the melting of glass except for the reaction with the body. In the first stage the glaze particles frit together and reduce the pore volume. The next step is the formation of a continuous glassy phase with the entrapment of bubbles left from the pores and the decomposition of the carbonates and clay. These bubbles work to the surface and break, where they form pits which soon smooth out. The forces bringing the bubbles to the surface are not gravity forces, as is the case in fining glass, but rather surface-tension forces attempting to bring the free surface area to a minimum.

The reaction between the glaze and body is an important one, as it forms an intermediate layer between the properties of the body and those of the glaze itself. This layer often contains mullite needles that grow from the body into the glaze, thus serving as anchors. The intermediate zone is better developed in cases where the body and glaze are fired together, since the interaction can be more complete.

Surface. It is desirable to obtain a smooth surface on a bright glaze. Superficially most bright glazes look smooth, but a careful examination under oblique illumination will show a number of shallow pits. In the case of some low-fired whiteware, the pits are so numerous that the gloss is definitely impaired, whereas the fine porcelain of Copenhagen has an almost perfect surface. Time and a nonporous body are apparently needed to obtain a perfect glaze surface.

Mat glazes develop, on cooling, fine crystals in the surface that break up the continuity and produce a sort of eggshell finish. These crystals are commonly anorthite ($CaAl_2Si_2O_8$), but mullite ($Si_2Al_6O_{13}$) and cristobalite (SiO_2) are often found. A good mat glaze usually contains lime and has a higher alumina content than a corresponding bright glaze.

Crystal development in the glaze. The fine crystals found in mat glazes are developed by normal firing schedules, for the number of crystals is so great that they cannot grow to a large size. Crystalline glazes have large crystals which grow to a diameter of as much as three inches under proper conditions (Fig. 15.4).

It has been shown that in growing crystals in glass it is first necessary to have a nucleus of a few atom groups as a starting point and then to cause this nucleus to grow. Certain glasses have separate temperature ranges for nuclear formation and for crystal growth, so means are at hand for controlling both the number and size of the crystals. This is illustrated in Fig. 15.5 for a zinc-titania glaze producing willemite (Zn_2SiO_4) crystals. The production of the crystals depends on heating the glaze above the nucleus-forming temperature long enough to dissolve all but a few of the nuclei, then dropping at once to the growing temperature, which is held until the desired crystal size is arrived at. The shape of the crystals may be controlled by the growth temperature, and their color by added transition metal oxides.

6. Fitting the Glaze to the Body

Cause of crazing. As was shown in the case of glass, glazes are very weak in tension but quite strong in compression. Therefore, if the glaze has a higher coefficient of

Fig. 15.4   Willemite crystal grown in a glaze (actual size).

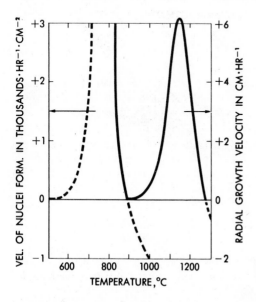

Fig. 15.5   Illustration of the formation of crystals from a glaze.

expansion than the body, on cooling it will go into tension and show the network of cracks known as crazing. The finer the network, in general, the greater the stress developed. On the other hand, very high compressive stresses may cause cracks known as peeling, a fault often occurring at edges and corners. Therefore, fitting a glaze to a body means more or less equalizing the coefficient of expansion of the two. However, the picture is complicated by the fact that the glaze as fixed in place may have a composition quite different from that computed from the formula, because of volatilization and solution of the body.

Measurement of glaze stresses. If a thin piece of body is glazed on one side only, it will curl under a stress developed in the glaze. The amount of this curl may be readily measured at any temperature in the following way. A tuning fork is made up of the body as shown in Fig. 15.6 and the outside of the prongs glazed. The whole is then fired to the normal maturing temperature and allowed to cool slowly

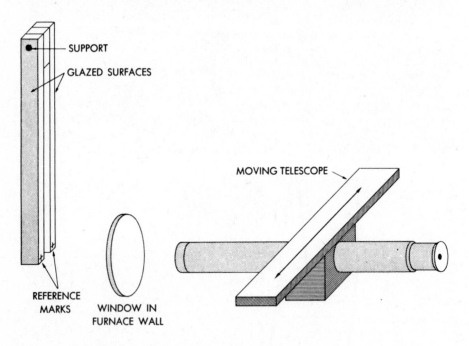

Fig. 15. 6   Method of measuring the stress in glazes.

in a furnace where the distance between the prongs may be measured with a telescope. If the prongs close up, the glaze is in compression and if they open up, it must be in tension.

Stresses in the glaze. Since glass, or the glaze, is strong in compression but comparatively weak under tension, glazes with any great degree of tension craze in a network of cracks, the pattern of which becomes finer as the tension becomes greater. Therefore, the desirable glaze will always be in slight compression because its coefficient of expansion is smaller than that of the body. The stress in a normal glaze is shown in Fig. 15.7. It will be seen that above 550°C there is no stress, as the glaze is too soft to support it. The final stress at room temperature *a* is a considerable compression, but this decreases in a few days to point *b* because of the delayed contraction in the glaze.

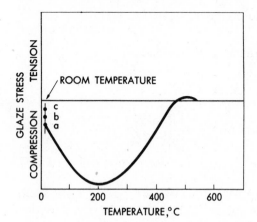

Fig. 15. 7   Stress in a glaze.

Crazing is a common defect in glazed ware, especially tiles. It can be controlled by careful design of glaze and body. There are many rules for preventing crazing by composition changes, but they do not always work. Only by a scientific approach through stress measurements can this trouble be effectively controlled. It has often been noted that mat glazes are less subject to crazing than bright glazes. This is because the fine crystals act as a reinforcement to the glass and thus permit it to carry considerable tension without crazing.

Delayed crazing. Even a well-fitted glaze may show crazing after exposure to moisture for a considerable period. This is brought about by a partial rehydration of the body, which causes a slight expansion and consequent increased glaze tension. This effect is most pronounced with porous bodies and is not found in the case of porcelain. Point *c* in Fig. 15.7 shows the stress after this test specimen had been exposed to moisture for some time. Had not the glaze been in high compression initially, crazing would have occurred.

The development of a good interface between the glaze and body is one of the best aids in preventing crazing. This is accomplished when the glaze and body are fired together.

7. Some Examples of Glazes

This section describes a few typical glazes.

Raw lead glaze. This type of glaze is simple to make up and apply. It is rather soft, scratching easily, but is brilliant and may be readily colored. The composition given below (from Binns) is quite satisfactory for a bright transparent glaze:

$$\begin{array}{l} 0.6 \ PbO \\ 0.3 \ CaO \ \cdot \ 0.2 \ Al_2O_3 \quad 1.6 \ SiO_2 \\ 0.1 \ K_2O \end{array}$$

This glaze may be made mat by increasing the alumina as follows:

$$\begin{array}{l} 0.50 \ PbO \\ 0.35 \ CaO \quad 0.35 \ Al_2O_3 \quad 1.55 \ SiO_2 \\ 0.15 \ K_2O \end{array}$$

To form an opaque enamel glaze, tin oxide is added:

$$\begin{array}{ll} 0.72 \ PbO & \\ 0.17 \ CaO \quad 0.17 \ Al_2O_3 & 1.93 \ SiO_2 \\ 0.11 \ ZnO & 0.33 \ SnO_2 \end{array}$$

All three of these glazes mature between $1050°C$ and $1120°C$ (cones 04 to 1).

Bristol glazes. These are raw glazes used on terra-cotta and stoneware where a lower maturing temperature than is possible with porcelain glazes is desired. A transparent glaze given by Wilson, maturing at $1175°C$ (cone 5), is quite satisfactory:

$$\begin{array}{l} 0.36 \ K_2O \\ 0.24 \ ZnO \quad 0.50 \ Al_2O_3 \quad 3.16 \ SiO_2 \\ 0.40 \ CaO \end{array}$$

A mat, opaque, Bristol glaze, also from Wilson, is made by increasing the lime and zinc oxide thus:

$$\begin{array}{l} 0.24 \ K_2O \\ 0.27 \ ZnO \quad 0.39 \ Al_2O_3 \quad 2.00 \ SiO_2 \\ 0.49 \ CaO \end{array}$$

Fritted glaze. These glazes are used for semivitreous and hotel chinaware. When properly made they have excellent working properties. Frits may be made in the laboratory by melting in clay crucibles, but it is more convenient to use commercial frits such as those listed by the companies making enamel frits.

A fritted glaze for a semivitreous body is given by Koenig (1951) as:

$$\begin{array}{ll} 0.43 \ CaO & \\ 0.26 \ PbO & \\ 0.12 \ K_2O \quad 0.27 \ Al_2O_3 & 2.60 \ SiO_2 \\ 0.06 \ Na_2O & 0.31 \ B_2O_3 \\ 0.13 \ ZnO & \end{array}$$

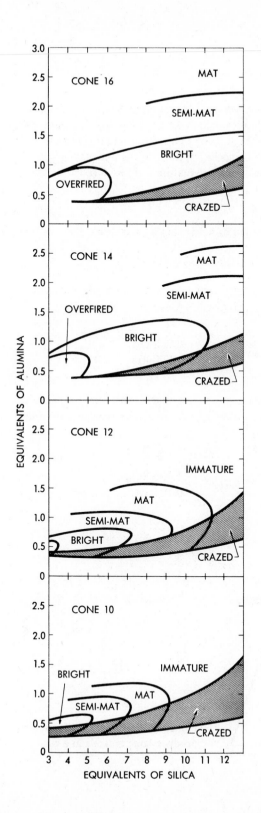

Fig. 15.8 Composition of the porcelain glazes. (RO = 0.3 K_2O, 0.7 CaO)

This matures at 1200 (cone 6).

Porcelain glaze. The field of porcelain glazes has been carefully studied by Stull and by Sortwell (1921). This type of glaze is easy to make up, nonpoisonous, hard, and has little tendency to craze. Only the high firing temperature required prevents its greater use. In Fig. 15.8 are shown the composition fields of the porcelain glazes.

A bright glaze may be made up as follows to mature at 1250°C (cone 9):

$$\begin{matrix} 0.30\ K_2O \\ 0.70\ CaO \end{matrix} \quad 0.58\ Al_2O_3 \quad 3.75\ SiO_2$$

A mat glaze maturing at the same temperature is made by increasing the alumina:

$$\begin{matrix} 0.3\ K_2O \\ 0.7\ CaO \end{matrix} \quad 0.65\ Al_2O_3 \quad 2.25\ SiO_2$$

By using multiple alkaline earths, the maturing temperature may be reduced. A bright glaze maturing at 1190°C (cone 6) is given by Geller and Creamer:

$$\begin{matrix} 0.217\ K_2O \\ 0.454\ CaO \\ 0.135\ BaO \\ 0.194\ MgO \end{matrix} \quad 0.352\ Al_2O_3 \quad \begin{matrix} 2.77\ SiO_2 \\ 0.82\ SnO_2 \end{matrix}$$

Crystalline glaze. Many types of glaze produce crystals with the proper heat treatment, as shown in Fig. 15.9. The following glaze produces excellent willemite crystals:

$$\begin{matrix} 0.235\ K_2O \\ 0.087\ CaO \\ 0.052\ Na_2O \\ 0.051\ BaO \\ 0.575\ ZnO \end{matrix} \quad 0.162\ Al_2O_3 \quad \begin{matrix} 1.700\ SiO_2 \\ 0.202\ TiO_2 \end{matrix}$$

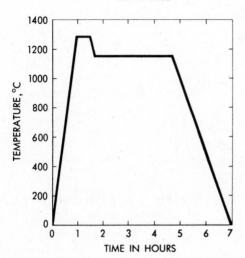

Fig. 15.9 Firing schedule used to produce large crystals in a glaze.

Reduction glaze. These glazes contain copper or iron oxide to produce reds and greens, respectively, under the proper firing conditions. These colors will be described in Chapter 17. It is also possible to produce these glazes by adding a reducing agent like ferrous iron or silicon carbide to the glaze or body and firing in an oxidizing atmosphere. The copper red glazes have been discussed by Mellor, who concludes that they are of colloidal origin and contain copper in the metallic state. However, there is evidence that crystals of cupric oxide may account for the color in some cases.

8. Glaze Defects

Ceramic glazes are plagued by many defects that can only be completely eliminated by careful control throughout the whole process.

Crawling. The glaze layer at maturing temperature is inherently unstable, as shown by Budworth (1971); nevertheless, it covers the body evenly if there is no break in its continuity. However, if there is a break in the layer, such as a drying crack, the glaze can draw back and expose a patch of the body (see Fig. 15.10). In extreme cases the whole glaze layer breaks up into beads as surface-tension forces tend to reduce the glaze surface to a minimum. On the other hand, some glazes (high-lead glazes) react so completely with the body surface that they will flow out and heal any small break.

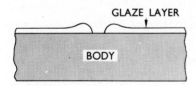

Fig. 15.10 Section of a glaze to show crawling.

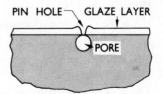

Fig. 15.11 Section of a glaze to show a pin hole.

    The solution to crawling is first to have the body surface free from dust or grease and a glaze layer with a low drying shrinkage. A high content of raw clay or over-fine grinding will inevitably cause crawling by producing a network of drying cracks in the glaze layer. It will also be noticed that glazes put on in a thick layer are more apt to crawl than those in thin layers, because of the more ready cracking when drying.

Crazing. This subject was discussed earlier in this chapter.

Pinholes. There are a number of causes of this defect and it is sometimes difficult to correct it. One cause is pores near the surface of the body, perhaps from air trapped in the casting slip. During firing the air in the pore expands and blows out a hole in the glaze (Fig. 15.11).

Orange-peel surface. This occurs if the glaze is applied by spraying and the droplets do not consolidate into a homogeneous layer. The condition can be corrected by applying the glaze more rapidly.

Specks. This is a common and annoying defect. It is caused by foreign particles either in the glaze slip or that fall on the glaze layer before or during firing. Microscopic and microprobe studies usually identify the specks and this leads to their source. Green specks (from copper) may come from valves in slip pumps or screens and brown and black specks often come from poor plant housekeeping.

Wavy surface. This defect is due to uneven application of the glaze, which causes surface irregularities too large to even themselves out in the firing.

References

Sortwell, H. H., High Fire Porcelain Glazes, *J. Am. Ceramic Soc.* **4**, 718, 1921

Mellor, J. W., Chemistry of Chinese Copper-red Glazes, *Trans. Ceramic Soc. (Eng.)* **35**, 364, 487, 1936

Norton, F. H., The Control of Crystalline Glazes, *J. Am. Ceramic Soc.* **20**, 217, 1937

Parmelee, C. W., *Ceramic Glazes*, Industrial Publications, Chicago, 1948

Koenig, J. H., and W. H. Earhart, *Abstracts of Ceramic Glazes*, College Offset Press, Philadelphia, 1951

Bloor, E. C., Glaze Composition, Glass Structural Theory and Its Application to Glazes, *Trans. Brit. Ceramic Soc.* **55**, 631, 1956

Edwards, H., and A. W. Norris, Examination and Maturing of Glazes, *Trans. Brit. Ceramic Soc.* **56**, 133, 1957

Lepie, M. P., and F. H. Norton, Life History of a Glaze Surface, *Trans. 7th Intern. Ceramic Congress*, 21, 1960

Reising, J. A., Zinc Oxide and Crawling of Glazes, *Bull. Am. Ceramic Soc.* **41**, 497, 1962

Franklin, C. E. L., The Maturing Behavior of Glazes. I. Bubble Development and Clearance, *Trans. Brit. Ceramic Soc.* **64**, 549, 1965

Franklin, C. E. L., The Maturing Behavior of Glazes. II. The Effect of the Ceramic Body, *Trans. Brit. Ceramic Soc.* **65**, 277, 1966

Budworth, D. W. A., A Theoretical Approach to Crawling of Glazes, *Trans. Brit. Ceramic Soc.* **70**, 57, 1971

Stull, R. T., Porcelain Glazes, *Trans. Am. Ceramic Soc.* **14**, 62, 1912

16

Enamels on metal

1. Introduction

Enamels form an excellent protective coating for metals. They are durable and washable, and may be made in white or various colors. The enamel-ware industry is primarily a metallurgical one, and buys the enamel in many cases from manufacturers who also supply the directions for application and firing. The industry is divided into three divisions: sheet-steel enameling, cast-iron enameling, and specialties such as aluminum signs and jewelry.

2. Low-Fusing Glasses

As the base metal on which the enamel is placed oxidizes and warps if heated to a high temperature, enamels must have a low softening point. Therefore, it is of interest to study the means of forming low-melting glasses.

The silica-oxygen network is a very stable one, but partial replacement of the silica with boron forms a portion of boron-oxygen triangles much weaker in bond strength than the tetrahedrons. Hence boron is an almost universal constituent of enamels. It is also possible to replace part of the oxygen by the weaker bonded fluorine in the glass network and thus give a lower softening glass; therefore many enamels contain fluorides.

The more active network modifiers produce a lower melting glass. For example, lithium may be used to replace sodium or potassium. The polarizable ions such as Pb^{++} weaken the network bonds as increasing amounts are added, and for this reason some enamels contain this element. Lead makes an easy-flowing, brilliant enamel, but for reasons of health it is generally limited to cast-iron enamels. Today, few lead-containing enamels are used.

At the same time that the bonds are weakened to provide greater fluidity, there is a tendency to develop less chemical resistance and also high thermal expansion. However, it is possible to obtain enamels with sufficient resistance by a careful selection and proportioning of the ingredients.

Enamel frits. Unlike most glazes, enamel contains almost 100% frit. The frit is melted in rotary furnaces or small glass tanks, quenched in water, and crushed.

Milling. The frit, together with mill additions such as clay and the opacifier, is wet ground in a ball mill to a definite fineness, perhaps 3% on a 200-mesh screen. The enamel slip is often aged 24 hours and is then ready for application. The method of enamel preparation is shown on the flow sheets of Figs. 16.6 and 16.7.

3. Theories of Adherence

One of the problems of enameling is to provide a good adherence between the enamel glass and the metal. A great deal of study has been given this problem, and as yet there is no generally accepted theory of the mechanism of adherence. It has been shown that a small amount of one of the transition elements is required. Some of the theories for adherence of enamel to steel are:

1. The transition element, preferably cobalt, promotes the growth of dendritic iron crystals from the base into the enamel. The crystals thus serve as anchors
2. Cobalt causes an adherent oxide coat on the iron base, into which the enamel fuses
3. The transition elements are polarizable and therefore promote chemical bonds between the metal and glass

4. Methods of Obtaining Opacity

In most cases it is desirable to produce a white or light-colored enamel. As the base coat is dark-colored because of the transition elements, it is usually necessary to cover this with an opaque coat. A high degree of opacity is desired, since the enamel thickness must be low for mechanical and cost reasons. Therefore, much work has been done in this field.

Theory of opacifiers. The opacity is obtained by distributing in the enamel glass small particles having a refractive index different from that of the glass. The scattering power increases as the difference in index between the two increases, and as the particle size approaches the wavelength of light.

The index of refraction of the enamel glass cannot be greatly varied and remains at about 1.5. However, particles of higher or lower index of refraction are available and are listed in Table 16.1.

Opacifiers added to the enamel. One of the usual methods of producing opacity is to add the finely divided particles to the frit in the mill. During the short time the enamel is heated, there is little chance for solution. Formerly SnO_2 was used for this purpose, but now the less expensive zirconium and titanium compounds have largely displaced it.

Opacifiers developed in the enamel glass. As was shown for the crystalline glazes, crystals may be grown in the glass at the proper temperature. The modern titania enamels are of this type.

Table 16.1   Properties of opacifiers

Material	Index of refraction	Theoretical relative scattering power
TiO_2	2.7	1.2
ZrO_2	2.2	0.7
Sb_2O_3	2.1	0.6
ZnO	2.0	0.5
SnO_2	2.0	0.5
$ZnAl_2O_4$	1.9	0.4
Al_2O_5	1.8	0.3
$MgAl_2O_4$	1.7	0.2
Enamel glass	1.5	0
CaF_2	1.4	0.1
NaF	1.3	0.2
Air	1.0	0.5

5. Jewelry Enamels

Base metals. These enamels are generally used on copper or copper alloys, but gold and silver are quite suitable. As the enamels are usually applied in color patterns, some means must be provided to keep one color from diffusing into the next. Three methods are used for this purpose. One, called *cloisonné*, outlines the pattern by fences of metal ribbon, hard soldered to the base as shown in (a) of Fig. 16.1. The enamel slip is then run into each separate compartment and fired (b). After cooling, the whole surface is polished down (c), leaving the partitions showing as fine metal lines between the colors.

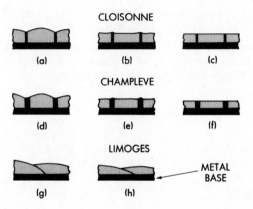

Fig. 16.1 Section through enameled jewelry to show the three common types of application.

In the second method, called *champlevé*, the dividing lines are applied in wax and the whole base is etched to take out the metal between the lines, which is later replaced by enamel as shown in (d), (e), and (f) of Fig. 16.1. In some cases the metal is cut out with an engraving tool.

The third method, often known as *Limoges*, applies the enamel directly without any partitions, as shown in (g) and (h) of Fig. 16.1, so that the colors diffuse into each other to some extent. Often several firings are employed before the piece is finished.

Composition of jewelry enamels. As these enamels must be fired at a lower temperature than those used for iron, they contain numerous fluxes to produce a low-melting glass. A typical formula for an enamel to be used on copper is:

SiO_2	20 parts
Red lead	61.5
Sodium nitrate	13.7
Boric acid	26.6

Frit, grind, and pass through a 100-mesh screen. Fire at 950°C for 2 minutes.

6. Enamels for Sheet Steel

Base metal. Enameling iron must have a low content of impurities in order to produce a good coating. The most generally used metal is Armco iron with the following analysis:

C	0.015%
Mn	0.020
P	0.005
S	0.025
Si	Trace

This sheet metal is shaped by pressing or drawing and must then be thoroughly cleaned before applying the enamel.

Ground coat. The ground coat is applied by spraying to a weight of about 2 ounces per square foot. A typical ground coat frit batch is shown in Table 16.2.

The sodium nitrate is an oxidizing agent and the transition elements are added to give adherence to the iron. This batch is melted in a frit furnace and quenched in water. It is then mixed with the following additions and put in a ball mill:

Frit	100 parts
Clay	7
Borax	0.5
Magnesium carbonate	0.12
Water	45

Table 16.2 Typical porcelain enamel frit batches

Constituent	Jewelry enamel	Sheet-steel ground coat	Sheet-steel cover coat	Cast-iron ground coat	Cast-iron cover coat (AR)	Aluminum enamel	Refractory enamel
Feldspar		29.5		30.4	8.7		
Borax		31.0		33.6	11.6		
Quartz	16.6	19.5	37.8	23.3	25.8	20.1	58.1
Soda ash		8.0				20.1	0.2
Sodium nitrate	10.2	5.0	7.9		14.3		
Fluorspar		5.3			5.1		
Cobalt oxide		0.7					
Manganese dioxide		1.0					
Dehydrated borax			16.4				
Silicon sodium fluoride					1.7		
Silicon potassium fluoride							
Lithium carbonate			2.5			8.1	
Titania			23.0		6.7	12.8	
Sodium antimonate					10.2		
Zirconium silicate				1.9			
Red lead	51.5			10.8	7.8	18.0	
Zinc oxide			1.0		5.1	0.8	4.6
Whiting			1.8		3.0		3.5
Boric oxide	21.7					3.6	5.1
Strontium carbonate						2.8	
Antimony pentoxide						2.9	0.3
Potassium nitrate			7.5			10.8	
Zirconium titanate							1.8
Barium carbonate							25.6
Potassium carbonate			2.1				0.5
Sodium aluminum fluoride							0.3
Mill addition	None	4–8% Clay Electro	4–6% Clay	2% Clay	None	1% Soda silicate	30% Al_2O_3

The clay, a special fine-grained material, is used for a suspending medium, but it also contributes to the opacity. The borax and magnesium carbonate are electrolytes to aid suspension. The grinding is carried on until about 70% remains on a 200-mesh screen. After application the layer is dried rapidly and then fired in a muffle furnace for about 2½ minutes at 830 to 885°C, the higher temperatures applying to the heavier gauge metals. Figure 16.2 shows the interface between the steel and ground coat, bubbles formed in the ground coat, and a portion of the cover coat.

Cover coat. Typical cover coats are shown in Table 16.2. These are fritted as before and the mill batch is:

Frit	100.0 parts
Clay	7.0
Borax	0.5
Water	50.0

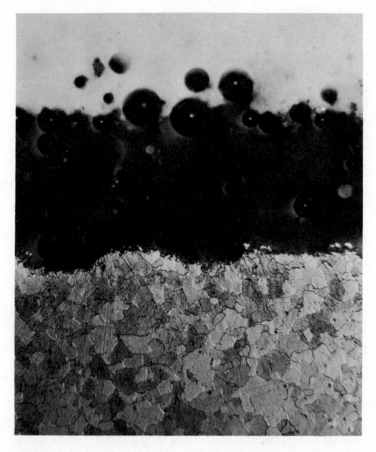

Fig. 16.2 Section of a two-coat titania enamel. The lowest section is the enameling iron. The black section above it is the ground coat, showing good adherence on the rough metal surface and numerous bubbles. The upper section is the white cover coat with some bubbles rising into it. 150 ×. (D.L.Gibbon and J.W.Smith, Ferro Corporation.)

This is ground, with 3 to 6% remaining on a 200-mesh screen. It is sprayed on the ground coat at a weight of about 6 ounces per square foot. The firing is for about 3 minutes at 800 to 850°C. For high-grade work two cover coats are used, the second one fired at about 10°C lower than the first.

The older tin oxide and antimony trioxide opacifiers have now been almost completely replaced by titania, which is in complete solution in the frit. On firing, crystals of anatase (TiO_2) and rutile (TiO_2) grow in the enamel to about 0.3 micron in diameter (Figs. 16.3 and 16.4). So efficient is this opacifier that comparatively thin cover coats may be used.

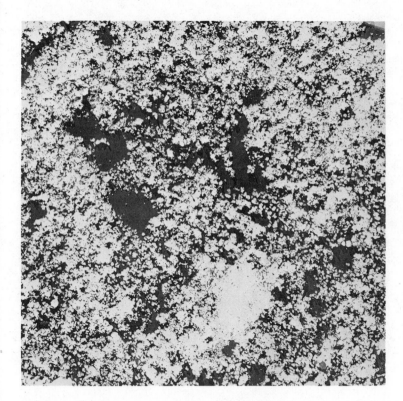

Fig. 16.3 Electron micrograph of anatase crystals in a titania-opacified cover-coat enamel. Carbon replica taken from the slightly etched surface. 20,000 X (Professor A. L. Friedberg, University of Illinois).

Some enamel is now made without a ground coat by treating the iron surface with a soluble nickel salt and applying the cover coat directly as shown in Fig. 16.5.

A typical flow sheet for a sheet-steel enameling is shown in Fig. 16.6.

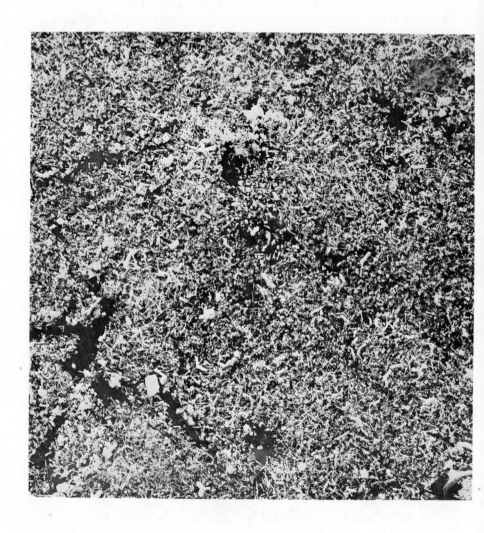

Fig. 16.4 Electron micrograph of rutile crystals in a titania-opacified enamel. Carbon replica taken from a slightly etched surface. 20,000 × (Professor A. L. Friedberg, University of Illinois).

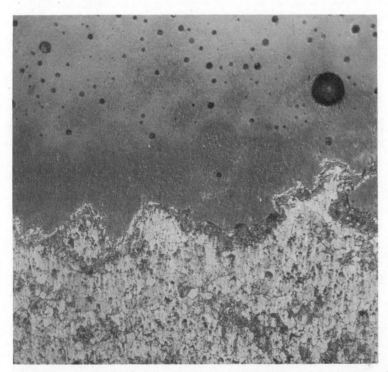

Fig. 16.5 Section taken 6° from the surface of a one-coat titania enamel (120 X). At the bottom of the picture is steel with small ferrite grains. Above this, along the jagged interface, is a dark layer from the nickel flash containing large rutile and $FeTiO_3$ crystals. Above this is the glass layer and at the top the titania-containing glass with numerous bubbles (Professor A. L. Friedberg, University of Illinois).

7. Cast-Iron Enamels

Base metal. The cast iron is the normal type of gray iron with a rather high silicon content to give fluidity and to prevent chilling in thin sections. A typical analysis is:

Carbon	3.25–3.60%
Silicon	2.25–3.00
Manganese	0.45–0.65
Phosphorus	0.60–0.75
Sulfur	0.05–0.10

The castings are cleaned up by grinding and sand blasting to give a bright surface.

Ground coat. A thin ground coat of the composition shown in Table 16.2 is then sprayed on at once.

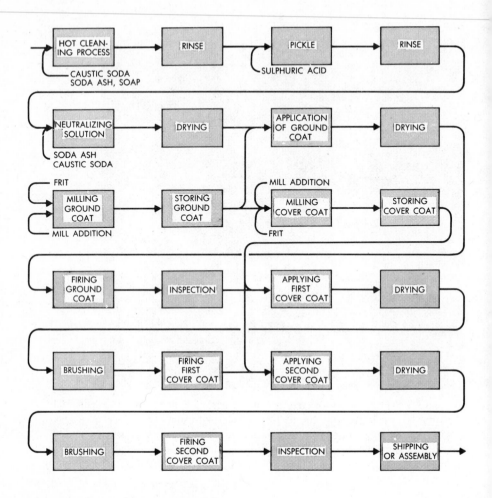

Fig. 16. 6   Flow sheet for sheet-steel enamel.

Cover coat. Although some small castings are coated by spraying as for sheet steel, the cover coat for larger pieces is generally applied as a dry powder to the heated iron. This process seems cumbersome, but produces excellent enamel surfaces. Since there is a large percentage of reclaimed powder to be remelted, the cast-iron enameler cannot economically buy the frit, but must smelt it himself. A typical composition is shown in Table 16.2.

The frit is dry ground and shaken onto the preheated piece, which is handled by lifting forks and placed on a manipulator so that it can be turned as the enamel is sifted onto it. This operation requires great skill and because of the radiant heat is

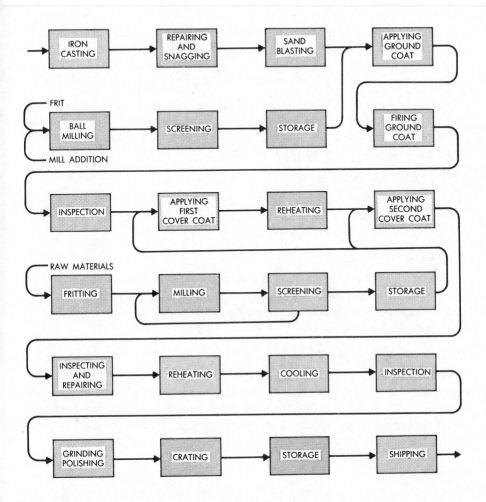

Fig. 16. 7   Flow sheet for dry-process cast-iron enamel.

a rather uncomfortable task. The manufacturers have been working on better handling equipment and automatic enameling machines. About one-third of the frit used falls on the floor and can be reclaimed. Because it picks up scale and other impurities, it cannot be used again immediately, but must be refritted. It is probable that methods could be developed to process this frit for direct reuse.

As the enamel powder falls on the hot iron, it fuses and sticks in place. After about $1/16$ in. has been built up, the piece is heated again, and a second coat put on which gives a total thickness of $3/32$ to $1/8$ in. A flow sheet for this process is shown in Fig. 16.7.

The opacifying agent used in dry-process cover coats at present is antimony trioxide or sodium antimonate. These partially or completely dissolve in the frit and recrystallize in the enamel fire. Great efforts have been made to replace this expensive opacifier with zirconia or titania, but without success.

Acid-resisting enamels. The regular cover coat is easily stained by fruit juices, so an increasing proportion of enamels are now made more stable by an increase in the silica content and a change in the fluxes. Lithia has been found very useful in these enamels. One example is shown in Table 16.2.

8. Enamels for Aluminum

In spite of the low melting temperature of aluminum (650°C), excellent enamels have been developed for this metal. A typical composition is given in Table 16.2. These enameled pieces are used for chalkboards and exterior building finishes.

9. Refractory Enamels

During World War II it was found possible to use refractory-coated carbon steel in some cases to replace the costly alloy steels. Since then there has been an increased use of these enamels on both carbon steel and ferrous alloys. The composition of one type is shown in Table 16.2.

10. Development of New Enamels

The development of a new enamel is a complicated process, since 10 to 30 ingredients must be proportioned to meet specifications for acid resistance, surface quality, and color and at the same time they must have the correct expansion properties, a maximum opacity, and a minimum cost. Progress has been made by some large companies in using the computer to handle this problem.

References

Andrews, A. I., *Porcelain Enamels,* 2nd ed., The Garrard Press, Champaign, Ill., 1961

Eppler, R. A., Reflectance of Titania Opacified Porcelain Enamels, *Bull. Am. Ceramic Soc.* **48**, 549, 1969

Eppler, R. A., and G. H. Spencer-Strong, Role of P_2O_5 in TiO_2-Opacified Porcelain Enamels, *J. Am. Ceramic Soc.* **52**, 263, 1969

Fisher, E. H., Electrostatic Enamelling, *J. Brit. Ceramic Soc.* **7**, 117, 1970

17

Decorative processes

1. Introduction

There are many different processes used for decorating ceramic ware. There is space here to cover only the more important or more striking ones, but the student who is interested will find much literature on the subject. Visits to our potteries will allow him to see mass-production methods of decoration in operation, while the European potteries may be counted on to illustrate the various hand-decoration methods.

2. Modeling

Probably the earliest type of pottery ornamentation was relief carving on the surface; at first as a simple scratched design and later as sculptured reliefs.

Relief decoration. A freshly thrown vessel can have a relief modeled on the surface directly with some of the same clay. Early Roman and modern Italian pieces are typical. Usually, however, the modeling is in the surface of a mold, in which a succession of pieces are formed by jiggering, molding, or casting. As an example, a plate with a raised border in relief may be taken. A model of the plate itself is turned from plaster, and a shallow groove is cut in the rim in the position of the relief band. This groove is filled with wax, in which the relief is roughly modeled. A negative plaster cast is then made of the positive surface of the plate and further modeling done in the plaster. Alternate positive and negative casts continue to be made, with modeling on each, until the relief is perfected. The reason for doing this is evident when one tries to model a delicate relief, for it is much easier to shape the humps than the hollows. The cross sections in Fig.17.1 will show the various steps in this process.

Applied reliefs. Early English sprig earthenware made in the middle of the seventeenth century produced relief decorations by forming wet clay in a small

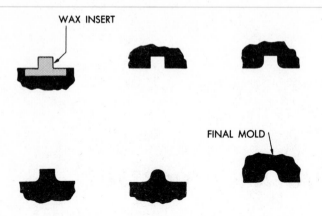

Fig. 17. 1   Steps in modeling a relief.

mold and at once pressing it, mold and all, onto the fresh surface of the vessel and then removing the mold. This rather crude method is not used at present, but the cameo ware, perhaps best known from the Wedgwood factory, is a refinement of this. Here the plastic body is pressed into small molds made of biscuit. After a short drying period the molded relief, often a very delicate structure, is gently burnished on the back with a spoon, which stretches the body slightly and allows it to be released from the mold; then it is temporarily set on a wet plaster block to prevent its drying. Now the surface of the piece is moistened with water and the relief section carefully applied. As the water dries, it draws the relief down into good contact. Obviously great skill is needed in this process.

An entirely different method of forming reliefs is the *pâte-sur-pâte* (paste on paste) method brought to a high degree of perfection in both England and France. As each piece is the individual work of the artist, the better pieces are in great demand by collectors. The process consists of building up a relief, usually in white porcelain slip on a colored background of raw, wet body. This is done with a brush, and layer after layer is added until a very low relief is completed. On firing, the varying translucency gives a most delicate effect, so that draperies may be superbly rendered. Often a clear glaze covers the whole. Although much inferior ware may be seen, a few artists have turned out exquisite pieces.

Sculpture in the round. Much sculpture has been produced in ceramic materials, from huge terra-cotta pieces to tiny figurines. In general, two methods are employed: the first is direct modeling in the body; the second consists of making a model, then a mold from it, and forming the body in the mold by pressing or casting. The direct method is generally confined to rather simple pieces where only one of a kind is needed. An example would be a portrait head. In this case the sections would be rather thick and it would be necessary to hollow out the center in order to dry and fire safely.

In most cases, however, production of a number of pieces of the same kind is necessary, so a model is made in wax, plasticine, or lead. If the design is complicated, it is cut up into an appropriate number of pieces and a plaster piece mold is made from each one. Each piece mold is then filled by casting or pressing and the units thus produced are assembled with slip into the pose of the original model. Some figure groups may have as many as 30 pieces that must be assembled with great precision. In the case of fine ware the parting lines are suppressed as described in Chapter 8. The biscuit porcelain of Sèvres and the Parian porcelain of England, made in the form of figurines, are excellent examples of this process.

3. Printing Methods

Early decorated pottery was hand painted and thus was too costly for the low-priced market. Therefore, the invention of Sadler and Green of Liverpool, England, in 1756, whereby they were able to apply a printed decoration, was most important. This process (and it is still used today for some types of ware) consisted of coating an engraved copper plate with a special ink containing a ceramic stain and transferring this ink to a thin paper which could then be placed on a piece of ware and rubbed down to transfer the ink again onto the body or glaze. The invention consisted fundamentally of transferring a design from a flat printing surface to a curved pottery surface by the intermediate step of a flexible membrane. This principle is inherent in most modern mass-production processes of ceramic decoration.

Today the Murray printing machine is widely used to apply decoration. This has a soft gelatine cone which is first pressed down on an engraved plate containing the printing medium and then pressed down on the ware to transfer the design.

Decalcomanias. This type of decoration was used extensively in the past because it is inexpensive and permits multiple colors. Although this method has been displaced to some extent by the silk-screen process, it is still important where several colors are desired.

The decalcomanias, or *decals*, are printed on special duplex paper — a heavy backing lightly attached to a thin tissue face that is coated on the outside with a soluble sizing — by the normal lithographic or offset printing process using up to eight colors. The inks are compounded of overglaze stains and a waterproof vehicle, such as a varnish, that hardens on the paper. In the dust process, varnish only is printed on the paper, and the dry ceramic stain is dusted onto it. The printing is done on large sheets, usually 45 inches by 29½ inches, containing all the patterns for one of each shape in a dinner set, which are then cut up to give a number of units for smooth application to the ware.

The decals are applied by first coating the glazed ware with varnish, letting this become tacky, and then rubbing the decal tissue with the backing paper removed, face down into this varnish. The ware is immersed in water, which floats off the decal tissue and leaves the ink in the varnish layer. The ware can then be fired like any overglaze decoration.

Silk-screen process. This process came into extensive use during the 1940s, to some extent displacing decals. It has proved particularly useful for decorating glassware.

The process uses a silk bolting cloth of 125 to 150 meshes per inch coated with a stencil film, except in those areas where a pattern is desired. In other words it is a reinforced stencil. The stencil may be prepared by cutting out areas of a thin decalcomania tissue mounted on a heavy backing. This tissue is mounted on the stretched silk screen with shellac and a hot iron, after which operation the backing is stripped off. If the design has fine detail, the photographic process is used, in which the stretched silk screen is coated with gelatine that has been sensitized with potassium bichromate. This is exposed through a negative to a strong light, and washed in water. The unexposed portions will be soluble and leave clear areas on the silk for printing.

In the actual printing operation the silk screen is placed a short distance above the surface of the ware. A rubber squeegee blade moves across the silk screen, forcing it against the ware, and at the same time pressing a viscous ink down through the holes in the open areas to form the pattern. Automatic machines have been developed to print by this method, but there is not space here to go into their design. If more than one color is required, each ink must be dried and a second printing made, registering with the first one.

4. Photographic Method

This method has many attractive possibilities, but as yet has not been used very extensively. Most of the processes consist of coating the ware with a light-sensitive, dichromate gelatine and exposing it through a negative. The unexposed portions are washed off, leaving a graded film which may contain coloring pigments to be fired on later.

5. Other Processes

Ground laying. This interesting process is used on fine china to provide an even color over large areas. The glazed piece is carefully painted with a solution of sugar and dye over all areas not to be covered by the ground. After drying, an oil (linseed oil and turpentine) is brushed on and then patted down with a soft pad to eliminate brush marks. Next, the dry, powdered overglaze color is dusted into the oil with a piece of cotton and the excess blown off. After drying about 24 hours, the piece is washed in water, which removes the areas painted with sugar solution. The pieces are then fired in the usual way for overglaze color and produce a ground that is uniform in both color and texture.

Hand decoration. Much fine ware is decorated by hand with brushes or needle sprays. This technique requires much experience and the European potteries have set up training schools for their apprentices. In this country little hand decoration is used, because of high labor costs, lack of trained personnel, and the reluctance of the American public to buy domestic production in the higher-priced classification.

Hand decoration may be applied under the glaze or over the glaze, as it is for the printing processes. The same ceramic colors are used, but the medium may differ. Some colors are applied with water-soluble gums or glycerine, others are applied with oils and a thinner like turpentine. A variety of brushes and application techniques are used which cannot be detailed here.

Much hand decoration is executed with the air brush, particularly for underglaze application. It is possible to spray on an area of color and then erase small spots to give a white design. For example, a Royal Danish Pottery fawn (Fig.17.2) is made by spraying on a brown underglaze, then with a pointed rubber erasing the color to form the white spots. The advantage of the spraying method is the very uniform surface produced.

Fig. 17. 2 Porcelain fawn with underglaze decoration (Royal Danish Pottery).

Stencil decoration. Simple designs may be readily achieved by spraying through stencils. These stencils are made of thin metal and are held a short distance from the ceramic surface. By proper registration of the stencils several colors may be applied to the same piece.

Stamping. Color may be applied with rubber stamps cut to the desired outline. Because of the distortion of the rubber, it is difficult to obtain clean, accurate application.

Sgraffito decoration. This is one of the simplest of methods. The green piece is coated with a thin layer of slip of a color different from the body. After the piece has dried, the outer coat is scraped away, to show the body in selected areas. The piece is then fired, and sometimes glazed.

Inlaying. Some of the early floor tiles in the English cathedrals have inlaid decorations that have withstood centuries of wear. This process is carried out by pressing colored clay into a mold so that the design is recessed below the surface. After the tile has dried slightly, another colored clay is pressed into this recess and scraped off flush. The whole is then fired to make a durable structure.

Encaustic decoration. This is a glaze decoration in enamel colors on tile. To separate the various colors so that they will not diffuse into each other, three methods are used. In the first, ridges are molded in the tile to define each area; in the second, grooves are used; and in the third, a narrow band of underglaze black. A heavy layer of glaze is flowed into each section by means of a syringe, and the whole fired. The effect is much like stained glass on a small scale.

Banding. Bands are put on plates by revolving them on a wheel and holding a brush against them at the proper place. This manual operation required great skill and long training. Now, banding machines have been designed to put on bands automatically at a great saving in labor costs.

6. Mechanism of Color Formation in Glasses

Properties of light. Visible light is a very narrow region in the long spectrum of electromagnetic waves that travel in the ether. The wavelength of the longest visible red waves is about 700 millimicrons and of the shortest visible violet about 400 millimicrons.

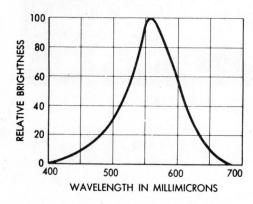

Fig. 17. 3 Color sensitivity of the human eye.

The normal human eye is a marvelous instrument evolved through millions of years from a rudimentary beginning in the skin of some aquatic animal. Color vision is not by any means the rule in the animal kingdom. Only man, the higher apes, birds, lizards, turtles, and fish are known to have it.

Although the color receptors of the human retina are not completely understood, it is known that there are responses to three regions of the spectrum, which combined give the effect of color. If either one or two of these response mechanisms are lacking, color blindness results. However, the sensitivity of even the normal eye is not equal over the whole length of the spectrum. It reaches a maximum in the yellow-green region, as shown by the curve in Fig.17.3.

Definition and measurement of color. Color to a physicist is a vibration in the ether; to a physiologist it is a stimulus to the retina; and to a chemist it is dye. The ceramicist looks at color in a relative sense in that he compares his product with some natural color, as the terms lilac, oxblood, or peach blow indicate.

The spectrum colors are defined by well-known terms and cover a wavelength range as follows:

Red	700–620 millimicrons
Orange	620–592
Yellow	592–578
Green	578–500
Blue	500–450
Violet	450–400

All colors except metallic lusters are the result of selective absorption of light transmitted through a transparent or translucent medium. A yellow glaze looks colored because light falling on it passes into the glaze layer and is reflected out again; in the process some of the blue and the red are absorbed, leaving the yellow to predominate in the emergent beam. It is well known that the apparent color of an object depends on the kind of light that illuminates it. A white paper looks red in red light, and the shadows on newly fallen snow look blue, for they are illuminated by the relatively weak, but bluish, light from the sky. It is a common experience to have two colors match well in daylight and to find later that they are quite different in artificial light.

There are two principal methods of measuring color. One is a method of comparison with a series of standard samples, for example, those supplied by the Munsell system. Here the different colors are arranged in the form of a color cylinder (Fig.17.4). The hues are distributed around the circumference with white, grays, and black on the axis. As this axis is approached, the colors become more gray or have less chroma. The colors lower on the cylinder are light or brilliant and the higher they go the darker they become. Each color has its complement directly opposite on the cylinder. Two colors are said to be complementary when they give a neutral shade on blending. This blending may be done on the familiar spinning disk, or by actual mixing. The color cylinder has around 800 numbered colors, so that a comparison may be readily made with a specific ceramic specimen and designated by number.

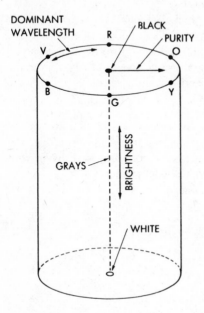

Fig. 17.4   Color cylinder.

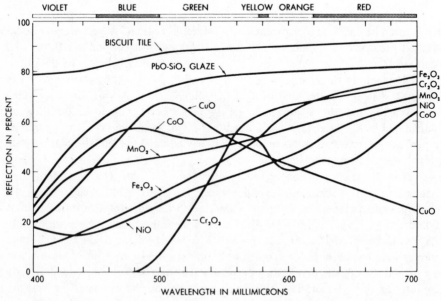

Fig. 17.5 Color curves for a simple lead glaze with various transition ions in solution.

As such color samples are not completely permanent, a more scientific method of measurement may be desirable. This can readily be done by means of the spectrophotometer, which records the transmission or reflection of each wavelength in the spectrum. In Fig.17.5 are shown the color curves of a glaze with various transition elements in solution. The curves may be converted into three color specifications expressed as dominant wavelength (hue), saturation, or the amount of white light mixed with monochromatic light of the dominant wavelength (chroma), and intensity (brilliance). For a more detailed discussion of color measurement the student should consult Hardy.

It should be kept clearly in mind that we have no method of measuring some of the more intangible factors connected with color. This is particularly true in ceramics, where translucency or transparency give a depth to color that is a large part of its charm. It is possible to find a Munsell color sample of ink on paper to match the color of a thick celadon glaze on an old Chinese vase, but the glaze has a depth that is quite absent from the ink.

Selective absorption by ions. Everyone is familiar with the fact that certain salts dissolved in water produce a colored solution. This is caused by selective light absorption by one of the ions. Ions are known to absorb light energy in three ways; first, by the vibration of the atom as a whole absorbing in the infrared region; second, by vibration of the electrons, absorbing in the ultraviolet region; and third, by orbit jumps, absorbing in the visible region. It is this last type of energy absorption that interests us here.

Coloring elements. Not all ions have the electron configuration that permits absorption in the visible range. Only those elements with an incomplete electron shell, such as the transition elements and the rare-earth elements, are capable of giving ionic absorption.

Color modifiers. A colored ion does not always give the same color, for the rate of electron vibration is influenced by the environment. For example, the valence state, the position in the glass network, and the type of ions surrounding it all influence the color of an ion.

Chromophore colors. There are certain cases where a color-absorbing complex is formed, much like organic dyestuffs, for example, in compounds of antimony and cadmium. These complexes are not stable at higher temperatures, but often give brilliant colors in enamels, low-temperature glazes, or overglaze decorations.

7. Solution Colors

Ions of the transition elements. The ions of the transition elements are shown in Table 17.1 with their probable colors, both when in the glass network and when in the modifier position. As will be seen, our information is not complete by any means, but a general picture may still be drawn.

Table 17.1 Colored ions in glasses

Ion	In glass network		In modifier position	
	Coordination number	Color	Coordination number	Color
Cr^{2+}		—		blue
Cr^{3+}		—	6	green
Cr^{6+}	4	yellow		—
Cu^{2+}	4	—	6	blue-green
Cu^+		—	8	colorless
Co^{2+}	4	blue-purple	6-8	pink
Ni^{2+}		purple	6-8	yellow-green
Mn^{2+}		colorless	8	weak orange
Mn^{3+}		purple	6	—
Fe^{2+}		—	6-8	blue-green
Fe^{3+}		deep brown	6	weak yellow
U^{6+}		orange	6-10	weak yellow
V^{3+}		—	6	green
V^{4+}		—	6	blue
V^{5+}	4	colorless		—

There are a few special cases that might be mentioned, such as the pure blue obtained from Cu^{++} in an alkaline glaze and the pink from Co^{++} in phosphate glasses.

Rare-earth ions. These ions differ from the transition ions in that the electron vibration causing absorption is not in the outer shells but in a more protected position inside. For this reason the vibration rate is not influenced by adjoining atoms, and the absorption spectrum is banded, rather than continuous as it is for the transition ions.

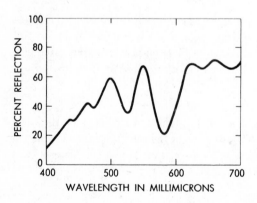

Fig. 17. 6   Reflection curve of a neodymium-containing glaze.

A rare-earth glaze containing Nd^{+++} ions gives a reflection spectrum as shown in Fig.17.6. The rare-earths do not generally give brilliant colors, but are used in filters because of their sharp cut-off and in some art glasses for their soft, dichromatic effects.

8. Colloidal Colors

Mechanism of colloidal color formation. A series of particles dispersed in a transparent medium with a different index of refraction scatters the light passing through. However, in colloidal colors we are concerned with much smaller particles. They are so small compared with the wavelength of light that they do not prevent complete transparency. Colloidal colors are most commonly produced by gold, silver, and copper dispersed in the metallic state as colloidal particles on the order of 50 millimicrons in diameter. The particles themselves have a selective absorption and a complementary selective reflection. As the particle size changes, there is a change in position of the absorption band, but as yet the complete mechanism of colloidal color is not understood.

Gold colloidal colors. Gold is dissolved in glass, not as the metal, as some have believed, but rather in the oxidized state as Au^+ ions which fit into the glass network in the same manner as K^+ or Na^+.

In some glass compositions, the Au^+ ion is stable enough to be frozen in on quick cooling so that the quenched glass is colorless. In other cases, especially with reducing agents, the Au^+ reduces to the metal even with quick cooling, and color is produced at once. Most gold ruby glasses, however, are formed by reheating the colorless, quenched glass for a sufficient period to decompose the Au^+ into metallic gold, thus:

$$3\ Au^+ = Au^{+++} + 2\ Au.$$

However, most gold rubies contain a small amount of SnO_2 or other variable valence oxide to decrease the solubility of gold by a reaction such as

$$2Au^+ + Sn^{++} = Sn^{++++} + 2Au.$$

The size of the metallic gold particles depends on the time and temperature of heat treatment, which in turn influences the hue. It is estimated that the gold content is about 0.0001 g/cc and the size is as follows:

Pink	4–10	millimicrons
Ruby	10–75	
Blue	75–110	
Livery brown	110–170	
Diffusion without color	400–700	

In Fig.17.7 is shown a section of a gold ruby greatly magnified to indicate the scale of the colloidal gold particles. The best gold rubies seem to occur in lead glasses. The composition $K_2O \cdot PbO \cdot 6SiO_2$ plus 0.0075% gold gives good results when reheated to 500 to 700°C. A composition of $Na_2O \cdot CaO \cdot 6SiO_2$, plus 1% Al_2O_3 and 0.0075% gold, if reheated to 650°C produces a good red. About 10% of the Na_2O should be derived from sodium nitrate to give an oxidizing condition in the melt. The addition of 0.5% SnO_2 also helps to form a good red.

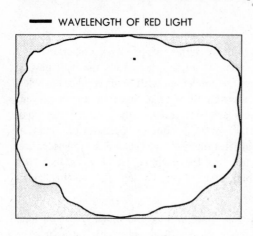

Fig. 17.7   Enlarged section of a gold ruby glass (7000 X).

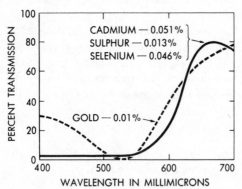

Fig. 17.8   Transmission curves of two types of red glass.

The transmission curve of a gold ruby is shown in Fig.17.8. The transmission in the blue end gives the characteristic purplish red to this glass.

Copper colloidal colors. The copper colloids are similar to those of gold. Cu^+ ions are in the network of the molten glass, but with a mild reduction on cooling form metallic copper as follows:

$$2\,Cu^+ = Cu^{++} + Cu,$$

or

$$2Cu^+ + Sn^{++} = Sn^{++++} + 2Cu.$$

The amount of copper needed is much greater than for gold, 0.1 to 0.5%. Therefore, the copper, if in the cupric form, will give a green. There is evidence that some of the red copper colors may be due to the red copper oxide, CuO, dispersed as crystals throughout the glass. The whole subject of reduced copper colors needs more study.

Silver colloidal colors. The silver ion Ag^+ is more stable than that of gold, so that 0.2 to 0.5% of this metal is required to cause precipitation. If reducing agents are present, a lower amount will suffice. The characteristic color of the silver colloid is yellow.

Cadmium-selenium colors. The important red glass used for signal lights is believed to be a colloidal dispersion of particles made up of a solid solution of CdS and CdSe. A transmission curve of this glass shown in Fig.17.8 indicates the almost complete lack of transmission in the blue end of the spectrum.

9. Colors in Crystals

Some glasses and glazes are colored by means of crystals dispersed throughout the mass. An important example is the scarlet glass made by the early Egyptians, which is colored by crystals of red copper oxide. Another common color is due to the bright red crystals of Pb_2CrO_6. The latter are unstable above $1000°C$ and transform to green Cr_2O_3 crystals.

Many other crystals may be colored by taking into solid solution one of the transition elements. For example, the colorless willemite crystals (Zn_2SiO_4) may be colored as follows:

Cu	Light green
Fe	Gray
Mn	Yellow
Cr	Green
Co	Intense blue

10. Ceramic Stains

There are many clays containing iron oxide or manganese oxide, which impart soft red, brown, or black color to the fired product. However, the majority of ceramic colors are derived from stains which may be colored glasses (smalts), but, more generally, are crystals colored by transition ions. These stains are used to color bodies and glazes or as under- or overglaze colors.

It is desirable that stain particles be relatively insoluble in the glaze; otherwise, the color will change. It has been found that crystals with the spinel lattice are relatively insoluble and also have the property of accepting other ions in the lattice even if their ionic radius varies quite widely from that of the host.

The unit cell of the normal spinel is shown in Fig.2.8, which gives the position of divalent and trivalent ions. Table A.1 gives the ions and their ionic radii. Other stains may be the transition oxide alone or this oxide combined with silica or some other oxide.

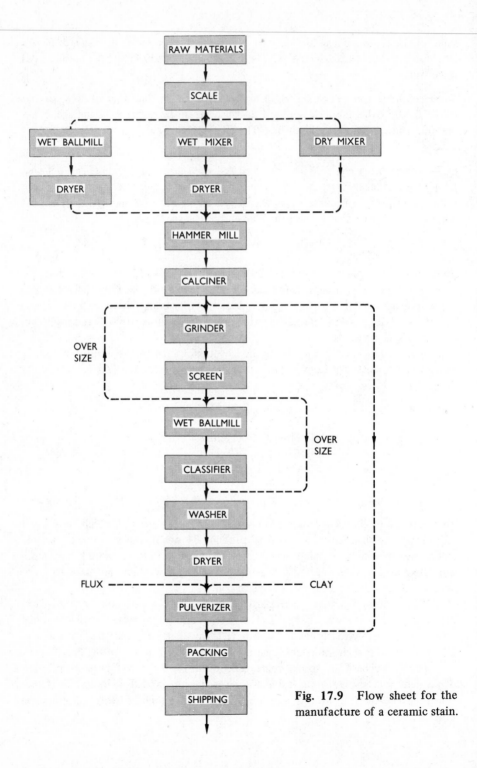

Fig. 17.9 Flow sheet for the manufacture of a ceramic stain.

It is well known that some stains are more stable than others at higher temperatures, so that at the low temperature of overglaze stains (850°C) many brilliant colors are available but at the temperature needed to fire porcelain glost (1500°C) only a few, rather soft colors can be found.

Stains may be used directly in the body and glaze, but for underglaze colors they are mixed with 5% flux and for overglaze colors 60 to 95% of flux is added to make the stain fuse into the glaze surface. Stains can be lightened by diluting with an inert white powder, such as ground bisquit, or they may have an opacifier added, often tin oxide.

In the past most potteries made their own stains, but the process is so specialized that they are now universally purchased from companies that make all the commonly used stains and are active in research to develop new ones. Perhaps the best way to explain the manufacturing process is by the flow sheet of Fig.17.9. One very important step is a thorough washing to eliminate all soluble material; otherwise, color will bleed out into the glaze.

Among the newer stains are the vanadium yellows made from vanadium-tin or vanadium-zirconium combinations, vanadium-zirconia-silica greens, and vanadium-zirconia-silica blues. If the reader wishes to know the compositions and stability of stains he should refer to Hainbach (1924) or Norton (1970).

11. Lusters

This type of decoration was used at an early date in Persia and by the Moors. It consists of a thin, more or less iridescent layer of metal or oxide on the surface of a glaze. There are two basic types, one produced in a reducing fire and the other under oxidizing conditions. Lusters may be either colored or colorless.

The luster is prepared by forming a metal resinate and mixing this with a vehicle, such as oil of lavender, for application to the glaze. The piece is then fired in a muffle at 600 to 900°C, whereby the resinate and vehicle are decomposed and the resulting carbon acts on the easily reducible oxide to produce the metal as a thin film on the glaze. It is probable that this operation takes place in the vapor phase, as the volatile bismuth is almost always present in a luster. Sometimes several lusters are applied one over the other to give special effects. For the reducing fire luster, the resinate is not used, but the metal salts, including bismuth, are applied with gum and water, and then fired in a muffle with a strongly reducing atmosphere.

12. Gilding

This is really an overglaze decoration consisting of a layer of a noble metal, usually gold, but sometimes silver or platinum.

Soluble gold process. In this process a soluble gold salt is incorporated in a varnish and applied to the glaze surface. On firing at about 700°C the gold is reduced to metal and deposited on the glaze in a thin layer in the manner of a luster. The gold, laid down in this way, is very thin and wears off quite easily so this method is used only on inexpensive ware.

Coin gold. In another method, powdered coin gold is mixed with a flux similar to those used for overglaze colors. Carried in a vehicle of oil or water-soluble gum, it is applied to the glazed piece. The firing is carried out at 700 to 800°C. After the firing the gold decoration looks brown with no metallic appearance. To form a polished gold surface the layer is burnished by rubbing with a smooth stone or, if a mat surface is required, with spun glass. The mechanical pressure causes the soft metal to flow into a continuous layer. This method of gilding is an expensive one and is only used on high-priced ware. The layer is quite durable, but being soft can be scratched.

References

Hainbach, R., *Pottery Decorating*, Scott Greenwood and Son, London, 1924

Searle, A. B., *An Encyclopedia of Ceramic Industries*, 3 vols., Benn, London, 1929–1930

Hardy, A. C., *Handbook of Colorimetry*, Technology Press, Cambridge, Mass., 1936

Peace, N., The Murray Printing Machine, *Trans. Brit. Ceramic Soc.* **57**, 527, 1958

Franklin, C. E. L., and J. A. Tindall, Organic Materials for Pottery Decoration, *J. Brit. Ceramic Soc.* **3**, 211, 1966

Grindey, W. T., and B. Mulroy, Synthetic Resins and Allied Materials in Pottery Manufacture, *J. Brit. Ceramic Soc.* **3**, 203, 1966

Shaw, K., *Ceramic Colors and Pottery Decoration*, Praeger, New York, 1968

Evans, W. D. J., Ceramic Pigments, A Structural Approach, *Trans. Brit. Ceramic Soc.* **67**, 397, 1968

Rado, P., *Introduction to the Technology of Pottery*, Pergamon Press, Elmsford, N.Y., 1969

Norton, F. H., *Fine Ceramics*, McGraw-Hill, New York, 1970

Demiray, T., *et al.*, Zircon-Vanadium Blue Pigment, *J. Am. Ceramic Soc.* **53**, 1, 1970

Eppler, R. A., Mechanism of Formation of Zircon Stains, *J. Am. Ceramic Soc.* **53**, 457, 1970

18

Fine ceramics

1. Introduction

This class of ceramics includes those having a body with a fine-grained structure. In most cases the body is white because only pure materials are used, but there are some exceptions.

2. Expressing Body Compositions

Batch formula. In the plant the body is expressed as the weights of the dry ingredients. A Sèvres porcelain body follows.

Materials	Weight	
Zettlitz kaolin	219.0 kg	69.5%
Quartz (potter's flint)	26.8	8.5
Feldspar (potash)	48.8	15.5
Marble ($CaCO_3$)	20.5	6.5
Total	315.1	100.0

This method is convenient for compounding, but it does not permit easy comparison of one body with another.

Mineralogical formula. As a means of comparing bodies, it has been customary in Europe to express the body composition in equivalent minerals as first suggested by Seger. For example, the raw materials in the body are mixtures of minerals which may be computed from the chemical analysis. While this method is not used to a great extent in this country, largely because of doubts about the validity of the calculation of the "clay substance", it is of considerable value in obtaining a broad view of body compositions.

The Sevres body given in the preceding section may be converted to the mineralogical formula if the following composition of the ingredients is known from the chemical analyses:

> Zettlitz kaolin, 95% clay substance, 5% quartz
> Quartz, 100% SiO_2
> Feldspar, 88% KNa spar, 2% Ca spar, 10% quartz
> Marble, 100% $CaCO_3$

Then a table is made up as follows.

Materials	Clay substance	Quartz	KNa spar	$CaCO_3$
69.5% Zettlitz kaolin	66.0	3.5	–	–
8.5% Quartz	–	8.5	–	–
15.5% Feldspar	–	1.6	13.6	0.3
6.5% Marble	–	–	–	6.5
Total	66.0	13.6	13.6	6.8

Equivalent formula. This mineralogical formula may be converted as follows into an equivalent formula.

Materials		SiO_2	Al_2O_3	KNaO	CaO
Clay substance	$\dfrac{66.0}{258} = 0.255$	0.510	0.255	–	–
Quartz	$\dfrac{13.6}{60.1} = 0.226$	0.226	–	–	–
Feldspar	$\dfrac{13.6}{54.7} = 0.025$	0.150	0.025	0.025	–
Calcium carbonate	$\dfrac{6.8}{100} = 0.068$	–	–	–	0.068
	Totals	0.886	0.280	0.025	0.068
Using Al_2O_3 = unity		3.16	1.00	0.089	0.24

The equivalent formula is then

$$
\begin{array}{c}
0.089 \text{ KNaO} \\
1.00 \text{ Al}_2\text{O}_3 \qquad 3.16 \text{ SiO}_2 \\
0.24 \text{ CaO}
\end{array}
$$

Chemical composition. The fired or dried body may be analyzed by the usual procedure. In the case of the dried body given in the preceding sections the results are as follows.

Silica	52.9%
Alumina	28.9
Ferric oxide	0.5
Lime	4.0
Magnesia	0.2
Potash	1.7
Soda	0.7
Water (combined)	9.1
Carbon dioxide	2.5
Total	100.5

This includes the minor constituents Fe_2O_3 and MgO now shown before.

3. Triaxial Bodies

These bodies are the traditional ones for whitewares of various types. They are called triaxial because they contain the three components, clay, quartz, and feldspar, although most of them have small percentages of other materials either as impurities or by design. The compositions of the most commonly used bodies are plotted in Fig.18.1. This triaxial plot is used to show relationships in mixtures of three components. Each corner of the triangle represents 100% of one component and any point in the enclosed area will correspond to 100% of the sum of the three components.

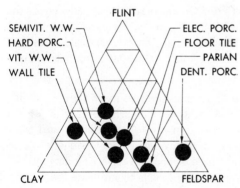

Fig. 18. 1   The composition of various triaxial whiteware bodies.

Semivitreous ware. This is the tableware commonly used in the United States. The body is generally white but may be colored. It is biscuit-fired at cone 7 to 10 (1210 to 1260°C), glazed with a fritted lead-containing glaze, and glost-fired to cone 1 to 5 (1140 to 1180°C). The decoration is mainly of the overglaze type put on by decals or silk screening. Typical compositions of bodies are shown in Table 18.1.

Table 18.1 Triaxial body compositions

Type of ware	Materials, %				
	Kaolin	Ball clay	Flint	Feldspar	Cone
Semivitreous ware	28	25	36	11	8
Hotel china*	37	8	35	20	11
Hard porcelain	46	–	34	20	15
Sanitary ware	30	10	28	32	9
Floor tile	32	–	10	58	10
High-tension insulators	15	30	20	35	10
Low-tension insulators	20	25	20	35	10
Dental porcelain	5	–	14	81	8
Parian porcelain	35	–	–	65	8

*1.5% of alkaline earth carbonates added.

This industry has been badly hurt by competition from Japan and Europe in spite of the 34% import duty.

Hotel china. This is a strong, durable ware made from a triaxial body with about 2% dolomite added to allow complete vitrification at cone 10 (1285°C). The glaze is matured in a second firing at cone 6 (1200°C). Decoration is usually under the glaze for durability. Recently, bodies with nearly double the usual strength have been made by the addition of alumina. A number of potteries make a thinner version of this ware, called Household China, for use in the home. Composition of a typical body is shown in Table 18.1.

Vitrified plumbing fixtures. Many items of vitrified ware are made for the building trade, but closet bowls, flush tanks, and lavatories comprise over 90% of the production. The body, as the name implies, is vitrified to a cream or buff color because low-cost raw materials are used. It is covered by an opaque white or colored glaze. Almost all the ware is once fired at cone 9 (1260°C) in large tunnel kilns. As the ware is slip cast much effort has been made to automate the process.

Hard porcelain. This is the material of the traditional tableware made in Europe and Japan and the chemical ware made in this country. The body is triaxial with the clay constituent entirely kaolin; the ball clays usual in American bodies are not employed as they reduce the whiteness and translucency. The biscuit fire is just high enough (1000°C) to make the ware safe to handle. It is then glazed and glost-fired at cone 14 (1400°C). In this country, hard porcelain is used mainly for chemical ware.

Floor tile. This is a triaxial body made by the dry-press process. Firing (there is usually no glaze) is carried out in fast-moving tunnel kilns at cone 10 (1290°C). The whole operation is well automated in the more modern potteries.

High-tension electrical porcelain. These very large pieces of porcelain are generally formed by extruding large cylinders of the plastic body, drying them to the leather-hard condition, and turning them on a lathe to exact size. A glaze is sprayed on and the insulator fired in a kiln to cone 10 (1290°C). Every precaution is taken to produce flawless pieces for the exacting service they are intended for.

Other triaxial bodies. Dental porcelains are very high in feldspar, as shown in Table 18.1, to give high translucency and a degree of self-glazing during the firing. The latter process is often carried out in a controlled atmosphere to reduce porosity to a minimum.

Parian porcelain is also high in feldspar to give the self-glazing properties often desired in small sculptures. Its composition is shown in Fig.18.1.

It has been found that some ceramic structures have properties that make them usable to replace bone in the human body. Strength and a porosity that will allow tissues to grow into the structure are necessary. Ceramic bodies have been found to possess advantages over metal for prosthesis of the hip joint, and tooth implants show interesting possibilities.

4. Bodies for Electrical and Electronic Uses

These bodies are used for a great variety of purposes, but in the space available only a rather brief treatment can be given.

Alumina. This strong nonporous oxide containing 93 to 96% alumina is mainly used for spark-plug cores. The dry mix is isostatically pressed in rubber molds, ground to size, glazed, and fired in small tunnel kilns. Its high electrical resistance and property of standing up under heat shock make it ideal for the purpose.

Another use is for printed circuit substrates of postage-stamp size and only a few tenths of a millimeter thick. Most of them are made by extruding or spreading with a doctor blade a thin sheet of fine alumina carried by an organic plasticizer. The dry sheet is punched to give the separate substrates, which are then fired on setters.

Many other electronic parts are of alumina, such as capacitors, cores, bushings, tube spacers, waveguide windows, and radomes.

Steatite bodies. Triaxial porcelains have served satisfactorily as insulators for 60-cycle use, but at high frequences losses become impossibly high. For this reason, the steatite body came into use, with a composition in the clinoensteatite field of $Al_2O_3 \cdot MgO \cdot SiO_2$ (Fig.18.2). The raw material is a pure talc, $(OH)_2 Mg_3(Si_2O_5)$, with some small additions such as $BaCO_3$. This body has a rather narrow firing range but can be successfully handled in modern kilns. The ware is generally pressed

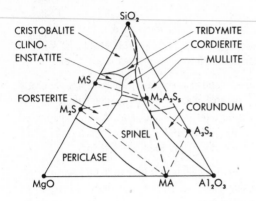

CRISTOBALITE
CLINO-
ENSTATITE

TRIDYMITE
CORDIERITE
MULLITE

SiO₂

MS

FORSTERITE
M₂S

M₂A₂S₅

CORUNDUM

SPINEL

A₃S₂

PERICLASE

MgO   MA   Al₂O₃

Fig. 18. 2 Three-component diagram of Al_2O_3-MgO-SiO_2. The primary fields are shown by solid lines and the compatibility triangles by dotted lines.

in steel dies with a low water content of 2% but with organic lubricants and binders. The finished piece, which may or not be glazed, has high strength and a low dielectric loss factor.

Ferroelectrics. The traditional dielectrics used in condensers had dielectric constants of around 6. When it was discovered that the alkaline earth titanates had dielectric constants of several thousands, entirely new possibilities for making minature circuits opened up. Barium titanate is the chief material, but other associated titanates are used. In order to get a uniform structure the titanates are usually made from coprecipitated colloids or fine grinding of constituents. Both high purity and exact stoichiometric proportions are desired.

The substrates are formed by several methods but flowing out a slip into a thin layer is often used. Firing is done in small electric tunnel kilns at 1330°C with a strictly oxydizing atmosphere. The wafers are set ten high on flat zirconia plates.

Ferromagnetics. The lodestone of the ancients (Fe_3O_4) was a nonmetal magnet, but little use was made of this until the 1940s. There are two classes of ferromagnetics, the soft magnets, and the hard or permanent magnets.

The soft magnet (ferrites) have the spinel structure

$$M^{++}Fe_2^{+++}O_4,$$

where M can be Mg^{++}, Ni^{++}, Co^{++}, Cd^{++}, Zn^{++}, or Mn^{++}. These materials are used in transformer cores, television deflection components, and computer memory cores. The raw materials are ground together, then pressed in molds. Firing is done in small tunnel kilns at temperatures of around 1300°C in a carefully controlled atmosphere. Temperature accuracy to ± 1°C is necessary.

The permanent magnets are of the magnetoplumbite structure, $Pb(FeMn)_{12}O_{19}$. The Pb^{++} may be replaced by Ba^{++} and Sr^{++}. These magnets find increased use every year. Every automobile contains 10 or 15 pounds of them and refrigerators have several for door closures.

Glass-bonded mica. Excellent insulators are made by fine mica flakes bonded by a glass frit. About equal amounts of fine mica and frit are milled together and pressed into a preform, which is heated to 700 to 1000°C to soften the frit. This preform is placed in a steel mold heated to around 400°C and compression- or transfer-molded in the same way as plastics. This gives a piece of extremely accurate size.

Electrically conducting ceramic bodies. Such bodies are of interest for resistors, heating elements, and electrodes. The best-known conducting body is made from silicon carbide and is commonly used for heating elements. For higher temperature use elements made of molybdenum disilicate can go up to 1700°C. Some oxides such as SnO_2, ZrO_2, and ThO_2 conduct well after they have been brought up to red heat. SnO_2 blocks are used as electrodes in contact with molten glass for heating some glass tanks.

5. Other Bodies

Bone china. This is the fine china of England. It has good strength, color, and translucency as well as attractive decoration, so it finds a market not only in all of Great Britain but also in the United States and Canada. The unique body consists of calcined bones, china clay, and Cornish stone. Production methods are much like those used in fabricating other pottery. The jigger has been largely replaced by the roller former for both flat and hollow ware. Much of the ware is still hand-decorated (Fig.18.3).

Frit porcelain. This fine ware is made in a few potteries such as Beleek in Ireland and Lenox in the United States. The body contains considerable glassy frit, which gives a high degree of translucency to the finished ware.

Wall tile. Most of the wall tile made in this country contains 60 to 80% talc. Dry-press forming is usual and the raw tile is glazed by spraying as it moves along on a belt. Firing is rapid (the complete cycle is less than an hour), at a maximum temperature of cone 03 to 1 (1090 to 1148°C). In modern tile plants production is highly automated. The majority of wall tiles are 4¼ in. square.

Art ware. Much low-priced art ware is made of a high-talc body fired at cone 06 (990°C). A low-temperature fritted glaze is used, often fired with the body. Higher-grade ware is often made of a fine stoneware body.

Low-expansion bodies. There has been an increasing demand for ware that can be safely heated in the oven and still be attractive on the table. Bodies containing 50% or more of corierite made from talc bodies have been fairly satisfactory but finding a good low-expansion glaze has been a problem. Bodies with still lower expansion coefficients may be found in the LiO_2-SiO_2-Al_2O_3 system illustrated in Fig.18.4. Both spodumene and eucryptite areas have zero or even negative expansion coefficients.

Glass-ceramic bodies containing lithia minerals have been very successful for ovenware. The glass is melted and molded in the usual way and devitrified under controlled temperature conditions to give the desired crystals.

Fig. 18. 3   Applying raised-gold decoration to bone china (Josiah Wedgwood and Sons).

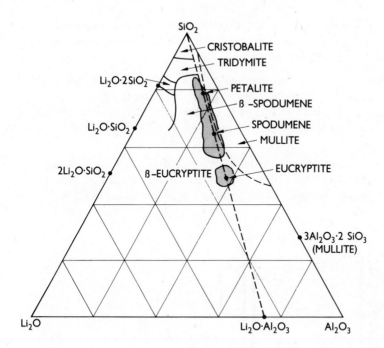

Fig. 18. 4 L_1O_2-SiO_2-Al_2O_3 diagram showing shaded areas of negative expansion.

Composites. In the last few years great efforts have been made to form composites of ceramic and metal or a combination of two or more ceramics, in an attempt to produce a stronger and tougher body. Often the reinforcing material has been in the form of fibers such as graphite, boron, or alumina. Although some promising materials have been made, as yet no commercial product has appeared.

Cutting tools. For many years metal-bonded tungsten carbide has been used for cutting tools and abrasion-resisting machine parts. Recently, sintered or hot-pressed alumina has shown excellent results as a cutting tool at high speeds, but for the best results machine tools of greater rigidity and power are needed.

Armor. It has been found that lightweight armor can be made of a sintered alumina face and an aluminum backing. This armor is being used for body protection and to cover the floor of helicopters.

Stoneware. A considerable quantity of stoneware is made for chemical purposes. This ware has been gradually improved by careful selection of clays and prereacted materials, and by aging and vacuum pugging. Large pieces can be made, such as 350-gallon crocks. The ware is fired at cone 10 to 12 (1280 to 1310°C). Some pieces are glazed, often with an Albany slip glaze.

Single crystals. Because large natural crystals are scarce and expensive much research has been applied to producing crystals in the laboratory. Several methods of crystal growing will be described briefly in this section.

1. Growth from a solvent. This method uses as solute aqueous solutions at controlled temperatures or molten fluxes such as cryolite. Reasonably large crystals of quartz, sapphire, titania, and garnets are made by this process. A variation uses solvents of water and other liquids at temperatures and pressures above the critical.

2. Melt freezing. In this method a melt in a crucible passes at a controlled rate from the hot zone to the cool zone with a seed crystal to initiate the action. Figure 18.5 shows the Stockbarger method, which produces large crystals of KCl, NaCl, and CaF_2.

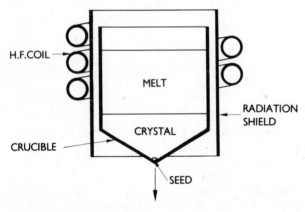

Fig. 18. 5 Melt-freeze (Stockbarger) method of single-crystal growing.

3. Withdrawal method (Czochralski). The melt is held in a crucible (Fig.18.6) and the crystal, started from a seed, is withdrawn continuously from the surface. Single crystal fibers may also be formed by this method.

4. Flame or plasma melting (Verneuil). This is the process used for many years to make single crystal boules of sapphire. The powder passes through the heating zone and builds up as droplets on the boule as it is slowly cooled (Fig.18.7).

5. Zone melting. In this method a slug of compressed powdered material is passed through a sharply confined hot zone where it melts and then recrystallizes (Fig.18.8). Since this method needs no crucible, there is no chance of contamination.

6. Vapor deposition. This method is used to form slender single crystals called whiskers.

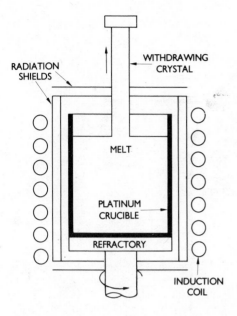

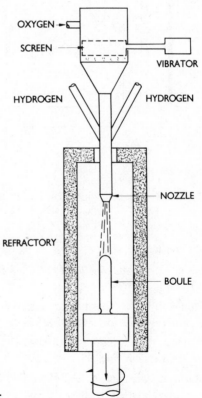

Fig. 18. 6   Drawing a single crystal from a melt (Czochralski method).

Fig. 18. 7   Growing a single crystal from flame-impelled powder (Verneuil method).

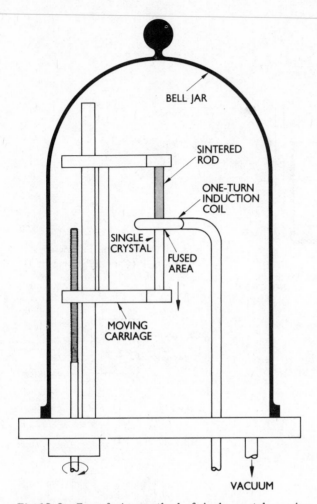

Fig. 18.8   Zone-fusion method of single-crystal growing.

References

Gould, R. E., *Making True Porcelain Dinnerware*, Industrial Publications, Chicago, 1947

Hummel, F. A., Thermal Expansion Properties of Some Synthetic Lithia Minerals, *J. Am. Ceramic Soc.* **34**, 235, 1951

Hummel, F. A., Significant Aspects of Certain Ternary Compound Solid Solutions, *J. Am. Ceramic Soc.* **35**, 64, 1952

Bush, E. A., and F. A. Hummel, High-Temperature Mechanical Properties of Ceramic Materials, II. Beta-Eucryptite, *J. Am. Ceramic Soc.* **42**, 388, 1959

Electronic and Newer Ceramics, Industrial Publications, Chicago, 1959

Clough, A. R. J., Plastic Forming by Roller Machine, *J. Brit. Ceramic Soc.* **1**, 391, 1964

Deri, M., *Ferroelectric Ceramics*, MacLaren, London, 1965

Completely Automatic Cup-making Plant, *Ceramics* **16**, 59, 1965

Mulroy, B., The Technical Control of Bone China, *Ceramics* **17**, 20, 1966

Heath, D. L., Electrical Porcelain Developments, *Bull. Am. Ceramic Soc.* **45**, 286, 1966

Fast Firing Wall Tile Body Investigation, *Ceramic Age* **83**, 7, 22, 1967

U.S.A. Standard Specification for Ceramic Tile, Tile Council of America, Inc., New York, 1967

Warshaw, S. I., and R. Seider, Comparison of Strength of Triaxial Porcelains Containing Alumina and Silica, *J. Am. Ceramic Soc.* **50**, 337, 1967

Stiglich, J. J., A Survey of Potential Ceramic Armor Materials, U. S. Gov. Doc. A.D. 666, 766, 1968

Hulbert, S. F., Use of Ceramics in Surgical Implants, Symp. Clemson Univ., Jan. 31, 1969

Tokar, M., Microstructure and Magnetic Properties of Lead Ferrite, *J. Am. Ceramic Soc.* **52**, 302, 1969

Hart, P. E., *et al.*, Densification Mechanism in Hot Pressing of Magnesia with a Fugitive Flux Liquid, *J. Am. Ceramic Soc.* **53**, 83, 1970

Bloor, E. C., Electrical Porcelain, I. A Review of Production Aspects, *J. Brit. Ceramic Soc.* **7**, 77, 1970

Maissel, L. I., and R. Glang (eds.), *Handbook of Thin Film Technology*, McGraw-Hill, New York, 1970

Jorgensen, P. J., and W. G. Schmidt, Final Stage Sintering of ThO_2, *J. Am. Ceramic Soc.* **53**, 24, 1970

Norton, F. H., *Fine Ceramics*, McGraw-Hill, New York, 1970

Lundlin, S. T., *Studies on Triaxial Whiteware Bodies*, Almquist and Wiksell, Stockholm, 1959

19

Refractories
and heat insulators

1. Introduction

The earliest refractories were sandstones or mica schist. These materials were used in the early iron forges, lime kilns, and glass furnaces in this country and later fire-clay brick were imported from England. Late in the 18th century firebrick were made in this country from local clays — the start of a great industry.

2. Heavy Refractories

These are the brick and shapes used for furnace linings. They have porosities of 10 to 25% and fusion points above 1400°C. The four main types are fire-clay, high-alumina, silica, and basic.

Fire-clay brick. These are made from plastic fire-clay and flint-clay or plastic clay and grog. There are ten classifications depending on the fusion point and firing temperature. Today nearly all of these bricks are formed by the dry-press process in heavy, automatic presses (see Fig. 8.13). The firing is done in tunnel kilns at temperatures of 1100 to 1500°C.

High-alumina brick. These brick have an alumina content running from 50 to 99%, which gives higher fusion points than for fire-clay brick and allows them to be used under more severe furnace conditions. These brick are made from diaspore or bauxite in the same manner as fire-clay brick except that the firing temperatures are usually higher.

The brick with AlO_3 content above 70% are made from chemically produced alumina grog. This highly abrasive mix may be pressed in a dry press with tungsten carbide dies giving a life of 75,000 cycles.

Silica brick. These are made from ganister rock with 2 or 3% of lime added. They are fired slowly, usually in periodic kilns. They are used in coke ovens and furnace roofs, but the demand for them has fallen off as no open-hearth steel furnaces are now being built. These brick are subject to spalling at the quartz inversion point, 573°C, but are quite stable above this temperature and maintain their strength close to their fusion point of 1700°C.

Basic brick. These comprise magnesite ($MgCO_3$) and magnesite-chrome mixtures. In this country we have little magnesite and almost no chromite, so the former is obtained from sea water and the latter imported. The magnesite is dead burned to drive off the CO_2 and sintered. The brick are formed by dry pressing and fired at around 1400°C. They are used in metallurgical furnaces where slag resistance is important.

Considerable amounts of magnesite and dolomite brick are made by bonding the grains together with hot tar. In other cases the fired brick is impregnated with hot tar. These brick give good service in oxygen steel converters.

Fig. 19. 1   Refractory setters used by potteries (Electro Refractories).

Sizes and shapes. The normal size for a straight brick is 9 by 4½ by 2½ in. but this is gradually being replaced by one 9 by 4½ by 3 in. A series of standard shapes are stocked by all manufacturers, details of which can be found in their catalogues. Other shapes, such as heavy kiln furniture, crucibles, and glass refractories, are stocked by a few makers. Then there are special shapes made to order, such as the setters shown in Fig.19.1.

3. Insulating Firebrick

In the early days of furnace construction, insulation, if used at all, was applied to the cool side of the furnace walls. Around the year 1930 insulation was gradually developed refractory enough to be used as the hot face of the structure. This had many advantages such as: (1) thin furnace walls with the same heat loss as the thick walls of heavy brick, (2) low heat capacity of the furnace wall thus allowing rapid heating and cooling of the structure, (3) lower fuel consumption of the furnace, and (4) lower floor area due to the thin walls.

The insulating firebrick (generally referred to as IFB) have the disadvantage of low resistance to fluid slags and to abrasion, and should never be used where these conditions are severe. However, they are very resistant to spalling (cracking from sudden temperature changes).

These firebrick are made from highly porosified fire clay, kaolin, high alumina clay, or alumina. The pores are usually made by adding wood flour to the plastic refractory material, molding and firing. The wood is burned out leaving fine, uniformly distributed pores.

The properties of these brick may be varied by changes in the raw materials and firing temperature. As the use temperature is raised the bricks become heavier and the thermal conductivity increases. Table 19.1 presents the properties of typical insulating firebrick.

Table 19.1 Properties of insulating firebrick

Type	Weight	$K*$ at mean temperature, ^{o}F				Use limit, ^{o}F
	pcf	500	1000	1500	2000	
16	18	0.6	0.8	1.5	—	1650
20	25	0.8	1.0	1.8	—	2000
23	30	1.0	1.2	2.0	2.3	2350
26	50	2.0	2.3	2.6	3.2	2600
28	55	2.3	2.5	2.8	3.4	2800
30	65	3.0	3.2	3.4	3.6	3050
33	85	4.0	4.5	5.0	6.0	3300

*K expressed in Btu/(hr) (ft^2) (°F) (in) (thermal conductivity)

4. Pure Oxide Refractories

A number of oxides with high fusion points are now produced in a pure, nonporous condition for high-temperature use.

Alumina. Sintered alumina is the most widely used of all oxide refractories. It is made from chemically produced material running from 92 to 99% Al_2O_3. The forming may be by slip casting, extrusion, dry pressing, or isostatic pressing, and the piece is then sintered at 1600 to 1900°C to produce a nonporous product. It is used in spark-plug cores, textile thread guides, wire-drawing dies, valve seats, substrates, and for many other purposes.

Beryllia. This oxide may be sintered into a dense form under much the same conditions as alumina, although elaborate precautions must be taken in handling the raw material, which carries health hazards. Because of the high thermal conductivity of the sintered material it is used in heat sinks for electronic circuits. It is also an excellent moderator for thermal nuclear reactors.

Magnesia. This may be sintered into a nonporous structure for use in crucibles and other small pieces, and it may also be sintered into a transparent form by the addition of 1% lithium fluoride, or by hot pressing without additions.

Silica. Silica sand can be fused into a dense mass in the electric furnace, ground, and mixed with water to form a casting slip. Objects of any size and complexity may be cast in plaster molds and fired. As there is little shrinkage the final pieces can be held to very close tolerances, and because of the low thermal expansion they withstand high thermal shocks. Above 1200°C there is a gradual devitrification.

Zirconia. The natural monoclinic zirconia transforms reversibly to the tetragonal form at about 1000°C with a large volume change, making it unsuitable for normal refractory use. However, if the zirconia is placed in solid solution with 4 to 12% of MgO, CaO, or Y_2O_3, it is held in the isometric form which has no transformation. Commercial zirconia refractories are stabilized with about 5% of MgO.

Zircon ($ZrSiO_4$). Dense zircon shapes are used in the paving of container-glass tanks and for nozzles in continuous casting of steel.

Mullite ($2Al_2O_3\ 3SiO_2$). Mullite is used for crucibles, pyrometer tubes, setters, and many other refractory uses.

Other oxides. Thoria, lanthia, yttria, tin oxide, and urania all have limited uses. ThO_2, with its high fusion point of 3300°C, is found useful in crucibles for metal melting. Lanthia has been used in some experimental refractories. Yttria (Y_2O_3), with 10% ThO_2 added, is calcined, ground, and cold pressed with an organic binder, then sintered at 1700°C to produce a transparent body for use as refractory windows and for infrared transmission.

Tin oxide is an excellent refractory, being very resistant to glasses and slags, but is difficult to sinter without additives. With proper activators (such as a combination of 0.2% CuO, 1% ZnO, and 1% Sb_2O_3 the room-temperature electrical resistivity can be as low as 1 ohm cm^{-3} and it can be used as an electrode in a glass tank.

The earlier nuclear reactors used metallic fuels but at present fuel elements of slightly enriched uranium oxide are generally used, since this allows a higher working temperature. The fuel element is often compacted UO_2 particles in a stainless-steel tube. In other cases 100-μ diameter UO_2 spheres are coated with pyrolitic graphite, all contained in a graphite matrix.

Some types of dense refractory are cast from electrically melted material. Glass tank blocks of alumina, mullite, and chrome-alumina are made, but the more recent block of 32 to 40% zirconia, 45 to 50% alumina, and the remainder silica is much used. Magnesia-chromite blocks are now made by fusion casting for use in steel making.

5. Nonoxide Refractory Bodies

Introduction. These bodies have been little employed by industry, as poor oxidation resistance and high cost have limited their use. However, since the nuclear and space age demands on refractories have greatly increased, their manufacture and properties have been reexamined. In this section some of the more interesting compounds will be described briefly.

Borides. The two most stable borides are TiB_2 and ZrB_2, which may be used in air up to 900°C. They have fusion points around 3000°C and a rather high thermal expansion of 7×10^6. No important use has been found for these borides.

Carbides. Silicon carbide (SiC) has a cubic structure at room temperature and changes at 2200°C to a hexagonal form, which decomposes at about 2750°C. This carbide is made in large quantities in the electric furnace (described in Chapter 20) for use in abrasive and heavy refractories. This is the most stable carbide in air, as a self-renewing film of protecting silica forms on the surface, allowing use up to 1700°C.

Silicon carbide refractories may be self-bonded by recrystallization at high temperature or bonded with clay or other material. A fine-grained, nonporous structure is made with high thermal conductivity and excellent heat-shock resistance.

Tungsten carbide (WC) and titanium carbide (TiC) are made in a dense, nonporous form which is bonded with cobalt to make cutting tools. Their hardness, 2500 to 3000, gives them excellent wear properties.

Uranium carbide, because of its high fusion point of 2450°C, has been used in nuclear fuel elements, but has now largely been replaced by the oxide. It is highly active in air, which makes handling difficult.

Boron carbide (B_4C) has a hardness a little above SiC and is used in small quantities for abrasion-resisting purposes, such as sandblast nozzles. It cannot be used in air above 450°C. It is fabricated by hot pressing in graphite molds.

Carbon and graphite. There are many known deposits of graphite, but few are satisfactory for refractory uses. Therefore, man-made graphite is produced from carbon by a long period of heating. These graphite blocks are machined easily, have high hot strength, high thermal and electrical conductivity, and excellent resistance to slags and nonferrous metals.

One of the largest uses of graphite is in electrodes for industrial furnaces, but there are many others such as nozzles for solid-fuel rockets.

Graphite may be vapor-deposited in the pyrolytic form as coatings or solids with interesting properties. It is highly anisotropic, so the thermal conductivity varies tremendously (see Table 19.2).

Graphite fibers are of interest because of their high hot strength and high elastic modulus.

Table 19.2 Pyrolytic graphite

Property	Direction	
	a	*b*
Density, g/cm^3	2.20	2.20
Thermal expansion, $^\circ C \times 10^{-6}$	0.9	27
Thermal conductivity, $cal/cm/cm^2/^\circ C/s$	0.6	0.005

*a, parallel to deposition surface;
  b, perpendicular to deposition surface.

Nitrides. Silicon nitride (Si_3N_4) is harder than silicon carbide and may be used in air up to 1400°C. It has been used in bonding refractories, and has possibilities for turbine blades and high-temperature bearings because of its high hot strength. Zirconium nitride, ZrN, is quite stable in air. Boron nitride, BN, can be produced by the following reaction:

$$Na_2B_4O_7 + 2N_2 + 7C = 2Na + 7CO + 4BN,$$

taking place as low as 1400°C. It is a white, soft, flaky powder that has been used as a high-temperature lubricant. It is reported that this nitride has also been obtained in a dense cubic form with a hardness equal to that of the diamond.

Sulfides. None of the refractory sulfides have found any general use. They are rather soft and not very stable in air. CeS is quite pyrophoric and must be handled with care.

Silicides. The only member of this group that is used to any extent is molybdenum disilicide, which is useful as a high-temperature, electrical-heater element. Thanks to the formation of a self-healing silica film on the surface it is stable in air up to 1700°C.

Manufacturing the nonoxide refractories. These materials may be formed by the usual ceramic processes, but the preferred way is to coat the fine starting powder with a wax binder and then press. Some of these fine powders are very active in air, in fact explosive, so a dry box is needed while handling them.

Firing should be done in a controlled atmosphere, for example, using ammonia for nitrides. As most of these materials are electrical conductors, high-frequency induction heating is convenient. Hot pressing has been found to give increased density and enhanced properties for some materials.

6. Refractory Plastics, Concretes, and Mortars

These materials are used in large quantities for furnace construction and repair. They must have good working properties and be sufficiently refractory for their particular use.

Plastics. These are mixtures of plastic fire-clay and grog, usually shipped and stored wet. They are used to build furnace walls, suspended arches, burner ports, and hearths. Often this material reduces labor costs of furnace construction, and if the temperatures are not extreme, gives excellent service.

Refractory concretes. These are made up of grog, fire clay, and 10 to 25% high-alumina cement. Mixed with water on the job they can be poured into forms like regular concrete, or they may be applied in layers with a cement gun. After setting they are dried and fired.

Mixtures of calcium aluminate cement and aluminous grog have excellent green and fired strengths. Other castables using aluminum phosphate as a bond have good high-temperature properties.

Refractory mortars. Brickwork must be laid with a mortar that will show little shrinkage on firing, adhere to the brick, and not fuse at the furnace temperature. Mixtures of clay and grog (about 20 mesh) which contain a small percentage of silicate of soda are found desirable.

7. Insulating Materials

Furnace insulation is made of many materials and comes in many forms. The commonly used materials are:

> Diatomaceous earth block
> Vermiculite block and nodules
> Slag-wool block, blanket, and bulk
> Glass-wool blanket and bulk
> Kaolin-wool blanket and bulk

Figure 19.2 shows an installation in the wall of a boiler furnace, and Table 19.1 gives the properties of the much-used insulating firebrick.

Fig. 19. 2 A flat-stud boiler furnace wall with insulation installed (Babcock & Wilcox Co.).

References

Batelle Memorial Institute, *Refractory Ceramics for Aerospace*, American Ceramic Society, Columbus, Ohio, 1964

Popper, P. (ed.), *Special Ceramics 4*, The British Ceramic Research Association, Stoke-on-Trent, 1968

Holden, A. N. (ed.), *High Temperature Nuclear Fuels*, Gordon and Breach, New York, 1968

Taeler, D. H., Tar-bonded, Tar-impregnated Basics, *Ceramic Age* **84**, 18, 1968

Norton, F. H., *Refractories*, 4th ed., McGraw-Hill, New York, 1968

Huggett, L. G., Design and Construction of Refractory Structures, *J. Brit. Ceramic Soc.* **6**, 123, 1969

Blakeley, J. D., Selection and Use of Thermal Insulating Materials in Furnace Construction, *J. Brit. Ceramic Soc.* **6**, 140, 1969

Kruger, O. L., and A. I. Kannoff (eds.), *Ceramic Nuclear Fuels*, American Ceramic Society, Columbus, Ohio, 1969

Alper, A. M. (ed.), *High Temperature Oxides: Part III, MgO, Al$_2$O$_3$, BeO, Ceramics*, Academic Press, New York, 1970

20

Abrasives

1. Introduction

Few people realize the important role played by abrasives in our present economy. The great industries, such as machine-tool, automobile, and aircraft, could not exist without grinding wheels, coated abrasives, honing stones, and polishing powders.

In recent years the abrasive industry has introduced new products and new applications of conventional products that have raised the performance capability far above that existing when the first edition of this book was published. New products of particular interest are man-made diamonds and the hard form of boron nitride. These innovations have been coupled with a new generation of grinding machines of such rigidity and power that both vitrified and organic-bonded wheels may be operated at higher and more efficient speeds than formerly.

It will be found that there is a scarcity of sound scientific and engineering literature covering the abrasive field in the English language, although there are numerous articles in German, Russian, and Japanese. One of the best sources of information is the issued patents.

2. Natural Abrasives

These are hard minerals found in nature, which were used as abrasives long before the electric furnace became available. Some of the more important ones will be mentioned in this section.

Emery. This is an impure corundum widely distributed on the earth's surface. In the nineteenth century several small emery mines were worked in the Berkshires in Western Massachusetts, and a number of associated grinding-wheel companies started up, which were the nucleus of the great abrasive industry today. Nowadays little emery is used in bonded abrasives.

Corundum. Massive crystals of this mineral were at one time imported from South Africa for use in snagging wheels.

Stones. The grindstone was formerly much used on farms and in shops for sharpening edged tools like scythes and axes, and is still employed today to some extent. It is composed of a fine, lightly consolidated sandstone. Whetstones of a finer structure are still used for putting a keen edge on carpenters' tools.

Feldspar. This is used in cleaners for glass, glazes, and enamels.

Garnet. This mineral is used for coated abrasives. It stands up about twice as long as quartz.

Quartz. This is used for sandpaper, in cleaning compounds, and as a loose abrasive. It is plentiful and inexpensive.

Diatomaceous earth. This siliceous material, in the form of microscopic skeletons of diatomes, is used in metal polishes. The most important deposits are in California.

Rouge. This finely divided ferric oxide, after calcining and classifying, is used for polishing glass, metals, and plastics. Other oxides used for this purpose are tin oxide and some of the rare earth oxides.

Diamonds. These extremely hard crystals, usually of the bort variety, come from the Congo and Siberia. This material is essential for cutting, grinding, and polishing hard materials.

3. Man-Made Abrasives

At the time when electric power became plentiful, at the start of this century, a whole new horizon opened up in the high-temperature field. Among the developments was the production of abrasive materials not found in nature in either sufficient quantity or purity to be useful.

Silicon carbide. This hard, brittle, and refractory crystal, SiC, has a complicated atomic structure which need not be discussed here. It is made in an electric furnace around 2200°C by vapor reaction between sand and coke:

$$SiO_2 + 3C = SiC + 2CO.$$

The furnace for producing this mineral is shown in Fig.20.1, a typical batch for which is:

Coke	16,000 lb
Sand	11,000 lb
Sawdust	1,800 lb
Salt	450 lb

The sawdust serves to open up the batch to make it more permeable to gases and the salt helps to remove iron as volatile ferric chloride. The furnace is run at a maximum of 3000 kw for 40 hours, by which time about 25% of the charge has been converted into usable SiC. Between 3 and 4 kwh are needed to produce one pound of product.

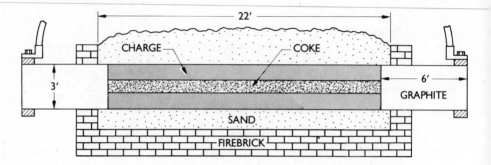

Fig. 20. 1   Cross section of electric furnace for producing silicon carbide.

The silicon carbide is made in several qualities, as:

Constituents	Crude	Black	Green
SiC	97.5	99.2	99.5
$Fe_2O_3 + Al_2O_3$	0.5	0.4	0.2
CaO	0.6	–	–
MgO	Trace	–	–
C	0.5	0.06	0.04
SiO_2	0.4	0.3	0.2

The raw lumps are crushed, de-ironed, and screened to provide closely sized fractions, and are sometimes washed and treated with HF to remove silica.

Fused alumina. This abrasive material is made by melting and purifying bauxite in an electric furnace. A cross section of a typical furnace is shown in Fig.20.2. The three-phase arc is started at the bottom and calcined bauxite is fed in to form a pool. The electrodes are slowly raised as feed is added until the melt reaches the top of the water-cooled shell. A voltage of 70 to 130 and a current of 1500 amperes is employed during a run of 48 hours to produce a pig of 45,000 pounds. It takes 1.1 to 1.4 kwh for each pound of Al_2O_3 produced.

Since bauxite contains Fe_2O_3 and SiO_2 as impurities it is fortunate that the highly reducing conditions during melting form ferrosilicon which settles and can be removed later.

There are variations of the furnace shown in Fig.20.2. For example, the water-cooled shell is sometimes replaced by a rammed carbon wall. Some furnaces have been built on a turntable to revolve slowly and give more even melting. Alumina can be melted in a continuous furnace in the same manner as calcium carbide, but the process is not suited to all grades of alumina.

There are several grades made, such as brown from bauxite with reduction and white from chemically prepared alumina. The composition of these grades are shown on following page .

Constituent	Brown	White
Al_2O_3	97	99.4
TiO_2	2	–
SiO_2	0.5	0.05
Fe_2O_3	0.5	0.05
Na_2O	–	0.5

Diamonds. After many years of unsuccessful attempts to convert carbon to diamond, this was finally accomplished by the combined efforts of scientists and engineers working with temperatures in the 1200 to 1600°C range and pressures up to 75,000 atmospheres. The crystals are generally quite small but careful tests show them to be equal to natural stones for industrial use.

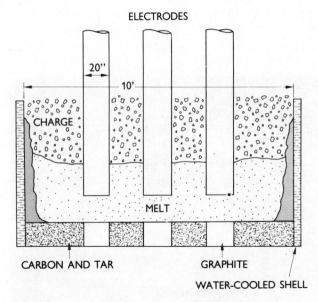

Fig. 20. 2   Electric furnace for melting alumina.

4. Grinding Wheels

In this country the grain (abrasive particles) is produced by three or four large manufacturers, although some is imported from Europe. Grinding wheels are made by twenty or thirty companies, two large ones and the rest quite small. The latter buy their grain on the open market.

Preparation of the grain. The sizing of the grain has been done by means of silk screens which have less tendency to blind than metal ones. Recently, screens of synthetic monofilament have come into use.

Forming the grinding wheels. Wheels are formed by bonding grains together with the following materials.

1. Ceramic (porcelain)
2. Ceramic (glassy)
3. Resins (thermo-setting)
4. Rubber
5. Silicate of soda
6. Magnesium oxychloride

The great majority of wheels are made with bonds 1, 2, and 3.

Ceramic-bonded wheels are molded either by a slip-casting method or by dry pressing in steel molds. After drying they are fired at around 1300 to 1400°C (Fig.20.3).

Fig. 20. 3   Abrasive wheels set on a tunnel-kiln car ready for firing (Richard Remmey Son Co.).

Resin(thermo-setting plastic)-bonded wheels are formed in a steel mold and then heat treated to set the resin.

The sizes of the wheels vary from a tiny dental tool to a huge wheel for making paper pulp up to 10 feet in diameter. The grain size and the hardness (amount of bond) vary in standard wheels over a wide range to suit particular purposes.

Although most grinding wheels are cylinders, many cup wheels (Fig.20.4) are produced, and form wheels, in which the surface is contoured to very close dimensions, are much used in industry. After some use the wheel may be trued to contour by a diamond dressing tool guided by a template. An interesting development is the periodic truing of the wheel by the pressure of a hard roller of tungsten carbide, boron carbide, or hardened steel, a process called crush dressing. A further development is the periodic truing of the grinding wheel by means of diamond truing rolls.

Grinding wheels with vitrified bonds are run at surface speeds of 6500 to 12,000 sfpm and resin-bonded wheels from 12,000 to 22,000 sfpm.

Diamond wheels or tools are of two types: the first has large stones (bort) held in metal retainers for such uses as core drilling, and the second, fine diamonds embedded in a metal or plastic bond for grinding and cut-off wheels or small drills.

Grinding wheels above 4-in. diameter are speed-tested at 50% above the operating rpm and wheels over 10-in. diameter are also balanced. There are some wheel failures, but in the author's experience all of them can be traced to improper use.

A safety code containing rules and regulations for the use, care, and protection of abrasive wheels has been issued by the American National Standards Institute, Inc. These rules and regulations should be followed rigorously when using abrasive wheels.

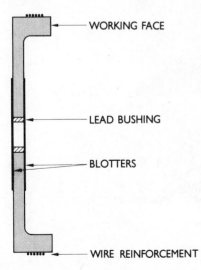

WORKING FACE

LEAD BUSHING

BLOTTERS

WIRE REINFORCEMENT   Fig. 20. 4   Cross section of a cup wheel.

The mechanism of grinding. The abrasive wheel removes metal or other materials by the action of the points and edges of the grains working as small but numerous cutting tools. A wheel, before use, is trued up by a diamond or other dressing tool, which shatters many of the surface grains to produce sharp edges, thus facilitating fast cutting. After a period of use these edges, in spite of their great hardness, become dull and the rate of metal removal decreases. However, if the grains are brittle or the bonding between grains is somewhat open, there will be fracture and pull out of the grains, and thus continuous exposure of fresh cutting surfaces. In this way it is possible to make a wheel that will continue to cut rapidly, but at the expense of wheel life. Therefore, a compromise between cutting rate and wear rate must be reached to give the best overall efficiency. This is done by varying the type, the size, and the shape of grain, and the type and amount of bond. Thus it is possible to furnish the user with the most efficient wheel structure for any given use.

The physical properties of the grain may be controlled by minor constituents, particularly TiO_2. Also the inherent crystal size is governed by the rate of cooling from the melt. It has been found that alumina may be toughened by the addition of another oxide in amounts up to 25%, for example, ZrO_2.

The cooled melt is crushed and screened to give variously sized fractions and passed through a magnetic separator to remove ferrosilicon. The microphotographs of Figs. 20.5 to 20.8 give a good idea of the microstructure of typical alumina grains.

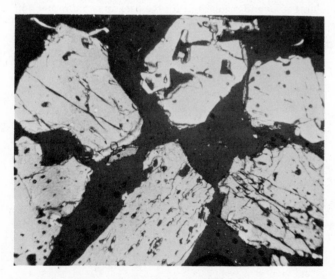

Fig. 20. 5   Fused calcined Bayer-process alumina.
Polarized light, 25 X (Carborundum Co.).

Fig. 20. 6   Regular bauxite-based fused alumina. Polarized light, 25 × (Carborundum Co.).

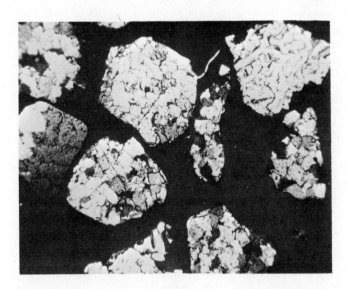

Fig. 20. 7   Chill-cast bauxite-based alumina. Polarized light, 25 × (Carborundum Co.).

Fig. 20. 8 Chill-cast zirconia-bearing alumina, containing 25% ZrO_2. 110 X (Carborundum Co.).

5. Coated Abrasives

We are all familiar with sandpaper, which has been in use for years, but today great quantities of coated paper and cloth are used in the form of belts or drums for grinding and polishing wood, metals, and plastics. The abrasive grains are garnet, alumina, or silicon carbide cemented to a paper or cloth backing with an adhesive. For polishing, crocus (iron oxide) is generally used.

6. Loose Abrasives

Vast quantities of abrasive powders in a vehicle such as water or oil are used for grinding, lapping, and polishing metals, ceramic articles, and plastics. Alumina, silicon carbide, diamond, and rouge are used in a variety of finenesses.

References

Coated Abrasives, McGraw-Hill, New York, 1958

Floyd, J. R., Alumina Abrasives – A Market Study, *Ceramic Age* **85** 36, 1969

McKee, R. L., Abrasives Industry – An Appraisal, *Abrasive Eng.* **15**, 30, 1969

A Grinding Wheel Every 90 Seconds, *Engineering* **206**, 201, 1968

Hargreaves, E. M., British Abrasive Research, *Abrasive Eng.* **15**, 36, 1969

Coes, L., Jr., *Abrasives*, Springer-Verlag, New York, 1973

21

Ceramic building materials

1. Introduction

Since the first settlers came to this country, brick have been made from the local clays in nearly every community for chimneys and fireplaces. In spite of the abundant supply of wood, most city buildings and some country homes were made entirely of local brick, a tradition brought over from England. Brick facings are still commonly used in many modern buildings, but other materials have greatly cut down on the use of brick.

2. Building Brick

There are many large, well-mechanized brickyards now taking the place of those small yards which still produce bricks by hand molding, sun drying, and firing by wood in a scove kiln. Tunnel kilns are used, but the scove kiln with movable scoving of insulating firebrick panels and mechanized setting and drawing is still used. Face bricks, with textured finishes and various colors produced by firing in controlled atmospheres or exposure to metal vapors, are generally made. It is of interest that the U.S. standard building brick is 8-in. long, derived from the Dutch brick around New York City. However, in old buildings we find the 9-in. English brick. There has been an attempt to make brick and other clay shapes to modular dimensions to simplify construction but it has not gained much ground as yet. Two factors have discouraged the use of brick in buildings: efflorescence and leaky walls. Both can be prevented by proper construction methods.

The absorption and the rate of absorption of building brick is an important factor, for if they are too high there may be freezing damage and the brick may draw the water out of the mortar too fast. On the other hand, if the absorption is too low, there will not be enough water interchange between brick and mortar to give a sound joint. It is believed an absorption value of 5% is about right.

Brick may be laid in beautiful patterns as can be seen in old walls in Europe. Figure 21.1 illustrates a few of the common bonds used in bricklaying.

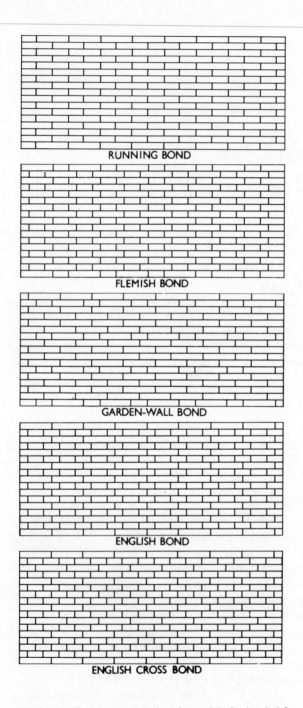

RUNNING BOND

FLEMISH BOND

GARDEN-WALL BOND

ENGLISH BOND

ENGLISH CROSS BOND

Fig. 21.1   Some common bonds used in laying brick.

3. Other Clay Products

Terra-cotta. This material is a large shape made from clay and grog and usually glazed. In the early part of this century many city buildings used this exterior finish; the McGraw-Hill building in New York City is a good example. Terra-cotta has a number of attractive features, such as a large range of surface colors and textures and a self-cleaning surface. Today little of this material is being used because of troubles with glaze crazing, frost damage, and a change in building styles.

Roofing tile. In the Western United States tile is a common roofing material, but little of it is seen in the East. Because of the colors available and its durability tile makes an excellent but rather expensive roof.

Quarry tile. These heavy tile, usually 8 in. square, of red or brown color, are used for outdoor and indoor terraces, hallways, etc. They make a pleasing and permanent floor.

Flue linings. These rectangular hollow tile are required by most building codes for brick chimneys. They are extruded from brick clay or shale.

Sewer pipe. Great quantities of this pipe are now made in standard sizes of 4- to 36-in. inside diameter with a laying length of 2½ feet. Most pipe is of a standard strength but a double-strength pipe is made for heavy loads. Formerly all sewer pipe was salt-glazed but a well-vitrified pipe was found to be not improved by the glaze, so a good portion is now left unglazed. Pipes larger than 36-in. may be laid up with segmented blocks.

Paving blocks. Many highways in the Middle West were paved with clay blocks. Properly laid on a concrete base, they make an excellent, long-lasting pavement, but because of cost they are not used as much as formerly. A common size is 3 by 4 by 8½ in. The brick race track at Indianapolis is a famous installation, put down in the year 1909.

Drain tile. This inexpensive, extruded tile is much used in land drainage.

4. Sand-Lime Brick

These brick are much used in Europe but in the United States they are less popular, being used mainly for back-up masonry. They are made from clean silica sand with 7 to 10% of hydrated or quicklime added and dry pressed. The brick are then placed in an autoclave at 150 to 200°C (115 to 270 psi steam) and held for 4 to 8 hours. A stable hydrocalcium silicate bond is formed, giving a crushing strength of 4000 to 12,000 psi.

5. Lime

Calcination. The raw materials for lime are limestone (Calcite, $CaCO_3$) or oyster shells (Aragonite, $CaCO_3$), which on calcination become quicklime (CaO). This reaction starts at about 900°C and is carried out in shaft or rotary kilns. The oxide is very reactive and may be slaked with water to form hydrated lime [Ca(OH)₂] with the evolution of considerable heat.

Mortars. The lime hydrate or lime putty is mixed with sand to form a mortar. This mixture slowly sets by absorbing CO_2 from the air to form $CaCO_3$, as well as by reacting with the silica to form calcium silicate. Lime mortars have been much used in the past for laying brick. In spite of their slow set and low mechanical strength, they formed excellent watertight walls, since their plasticity produced well-filled joints. At present, mortar is made by adding Portland cement to the mixture for more rapid set and greater strength. A typical mortar for brick setting above grade would be:

> 1 volume Portland cement
> 2 volumes lime hydrate
> 8 volumes sand

6. Portland Cement

Portland cement is one of the hydraulic cements that sets in the presence of water. It is one of our most important structural materials, having good compressive strength and reasonable durability. However, it is a very basic material and, from the point of view of the geologist, is not an ideal composition to withstand weathering.

Manufacture. Portland cement has a composition as follows:

Silica	19–25%
Alumina	5–9
Ferric oxide	2–4
Lime	60–64
Magnesia	1–4
Sulfur trioxide	1–2

In Fig.21.2 is shown a three-component diagram of lime-silica-alumina with the area of Portland cement composition indicated, when ferric oxide is combined with the alumina, and magnesia with the lime. On the same diagram is shown the area for high-alumina cements, which will be discussed later.

The composition for Portland cement may be obtained by combining clays and limestone in the correct proportion, or it may be made of corrected slags from the steel industry. There are two processes used, the wet and the dry. In the former the raw materials are wet ground in ball mills to a slurry for feeding the rotary kiln. In the latter the grinding is dry and the powder is fed into the kiln.

The burning is carried out in large rotary kilns with inside diameters up to 10 feet and lengths up to 250 feet. As the raw material slowly moves down the revolving kiln, it encounters hot gases moving in the opposite direction and so is heated up to around 1600°C when it reaches the hot end. It is then discharged in the form of clinker nodules. The chief reactions occurring in firing are shown in

Fig.21.3. The firing is usually accomplished by means of a powdered coal burner, and the waste heat is recovered in a boiler at the other end. The kiln is lined with refractory material, dense fire clay at the cool end and either a high-alumina brick or a magnesite brick at the hot end. Good operation depends on building a layer of clinker on the brick surface. A cross section of a kiln is shown in Fig.21.4. The clinker is cooled in a rotary cooler below the kiln with recovery of the heated air for the burner. The clinker is then ground in ball mills and bagged. A flow sheet for the dry process is shown in Fig.21.5.

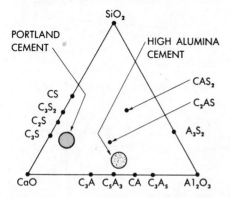

Fig. 21. 2   Composition of Portland cement in the $CaO-Al_2O_3$ system in weight percent. The composition of high-alumina cement is also shown.

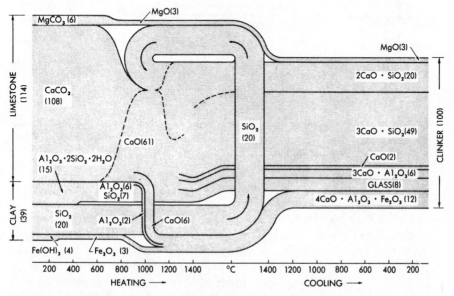

Fig. 21. 3   Reaction occurring during Portland cement clinker formation.

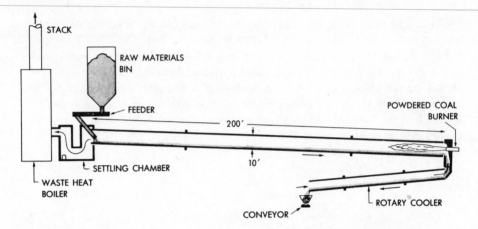

Fig. 21. 4 Cross section of a rotary kiln for burning Portland cement clinker.

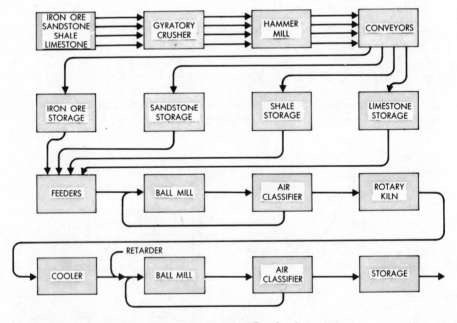

Fig. 21. 5 Flow sheet for making dry-process Portland cement.

Mechanism of setting. Portland cement contains largely the minerals tricalcium silicate, dicalcium silicate, tetracalcium aluminum ferrite, and tricalcium aluminate. On hydration, there is solution, recrystallization, and the precipitation of colloidal silica that causes setting. As Portland cement would ordinarily set too rapidly, a retarder of 3% gypsum is used. This reacts with the tricalcium aluminate to form calcium sulfoaluminate that slows the set and decreases the shrinkage.

Portland cement develops heat in setting, which in massive structures may cause dangerously high temperatures unless cooling is used. The heat of setting for some of the minerals is:

$3CaO \cdot Al_2O_3$	207 cal/gm
$3CaO \cdot SiO_2$	120
$4CaO \cdot Al_2O_3 \cdot Fe_2O_3$	100
$2CaO \cdot SiO_2$	62

Special types of Portland cement. High early strength cements are produced by an increase in tricalcium silicate and extra fine grinding. Low-heat cements are made by eliminating most of the tricalcium aluminate. Sulfur-resisting cements are also low in tricalcium aluminate.

7. High-Alumina Cement

This cement has the following composition:

CaO	35–42%
Al_2O_3	38–48
Si_2O	3–11
Fe_2O_3	2–15

This composition is given on the diagram of Fig.21.2 to show its relation to Portland cement. The manufacture is the same as for Portland cement except that the clinker is more nearly fused due to its lower softening point. Also, no retarder is used.

The setting is caused by the formation of hydrated alumina from tricalcium aluminate. This type of cement gains the same strength in 24 hours that is attained by Portland cement in 30 days, and is therefore used where quick setting is helpful. It is also more resistant to sea water and shows less fluxing in refractory air-setting mortars.

8. Gypsum Plaster

This material, often called plaster of Paris, is not only important to the ceramic industry as a mold material, but is also used extensively as a building material in blocks, wallboards, and plasters.

Raw material. The raw gypsum rock, when pure, consists of hydrated calcium sulfate, $CaSO_4 \cdot 2H_2O$. There are deposits containing 99% gypsum, but others may drop as low as 65% with impurities of limestone, quartz, or shale.

Calcination. The rock is fine ground and then calcined in iron kettles at a temperature of 160 to 175°C. Most of the plaster is the so-called first settle, that is, carried over the point where the first gases have escaped. In this condition the composition is $CaSO_4 \cdot \frac{1}{2}H_2O$, the so-called hemihydride. In a few cases, the calcination is carried further to produce anhydride or $CaSO_4$. A flow sheet of the kettle process is shown in Fig.21.6.

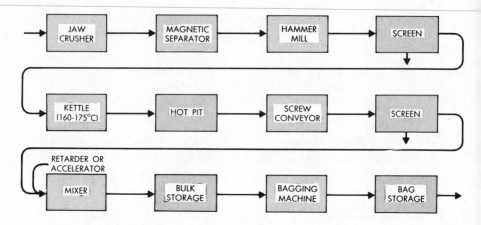

Fig. 21. 6   Flow sheet for making gypsum plaster.

Mechanism of setting. When the hemihydride is mixed with water there is a partial solution and a recrystallization of gypsum into a hard mass with some evolution of heat. The properties of the set material are influenced to a considerable extent by the amount of water added, as shown in Fig.21.7. Accelerators or retarders may be used but ordinarily they are not needed, as the normal time of set is about 20 minutes. This is decreased by longer mixing or by using seed crystals of previously set plaster.

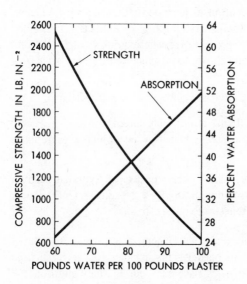

Fig. 21. 7   Properties of gypsum plaster with varying amounts of water.

Special types of plaster. Hard gypsum plasters are made with greater strength and hardness than ordinary potters' plaster. These are useful for models and pressing molds.

9. Oxychloride Cements

Magnesium oxychloride cements, often called Sorel cements, are made by mixing caustic MgO with a $MgCl_2$ solution. Magnesium oxychloride is slowly formed to produce a hard structure. These cements are used for interior floors. Zinc oxychloride cement was used for dental purposes but has been displaced by other types.

10. Silicate Cements

These cements are made from silicate of soda and a filler such as fine-ground quartz. On drying, the mass becomes hard, and on heating slowly it may be stabilized. These cements are used for acid-resisting construction, since acid precipitates silica gel, which is then inert.

Silicate of soda with alumina and a fluoride have been used for dental purposes. Silica and alumina gels are formed with the valuable property of translucency.

11. Phosphate Cements

These cements have been used for dental purposes. The oxides of Zn, Zr, or Cu are mixed with phosphoric acid, which reacts to form the metaphosphates.

For air-setting refractory purposes the phosphate cements look promising because of their strength at all temperatures.

References

Voss, W. C., Classification of Brick by Water Absorption, *Ind. and Eng. Chem.* **27**, 1021, 1935

Akroyd, T. N. W., *Concrete: Properties and Manufacture*, Pergamon, London, 1962

Orchard, D. F., *Concrete Technology*, 2 vols., Wiley, New York, 1962

Witt, J. C., *Portland Cement Technology*, Chemical Publishing Co., New York, 1966

Holdridge, D. A., and E. G. Walker, Dehydration of Gypsum and the Rehydration of Plaster, *Trans. Brit. Ceramic Soc.* **66**, 485, 1967

Midgley, H. G., The Mineralogy of Set High-Alumina Cement, *Trans. Brit. Ceramic Soc.* **66**, 161, 1967

Ratcliffe, S. W., D. Hall, and W. L. German, Glazes for Sewer Pipes, *Trans. Brit. Ceramic Soc.* **67**, 69, 1968

Svec, J. J., Trends in United States Brick Production, *11th International Ceramic Congress,* London 1968

Yount, J. G., Jr., and W. H. Powers, Simulated Service Tests of Cement Kiln Brick, *Bull. Am. Ceramic Soc.* **48**, 716, 1969

Shaw, K., Updating Clay Product Drying Processes, Part 1, *Brit. Clayworker* 78, 38, 1969

Hanna, K. M., and S. A. El-Hemaly, Effect of Curing Conditions on the Total Water Content of Hardening Portland Cement Paste, *J. Appl. Chem.* **19**, 241, 1969

Lea, F. M., *The Chemistry of Cement and Concrete*, Chemical Publishing Co., New York, 1970

Houseman, J. E., and C. J. Koenig, Influence of Kiln Atmosphere on Firing Structural Clay Products, *J. Am. Ceramic Soc.* I, **54**, 75, 1971; II, **54**, 82, 1971

Peray, K., and J. J. Waddell, *The Rotary Cement Kiln*, Chemical Publishing Co., New York, 1971

22

Glass products

1. Introduction

The glass industry has been progressive in developing and marketing a great variety of products. In this chapter there is space to cover only the more important items of a wide array.

2. Sheet Glass

This product, often called "window glass," is used in window sashes, picture frames, greenhouses, and for many other purposes. It comes in single and double thicknesses and many sizes. Anyone who compares modern window-panes with those in houses fifty or more years old will see a marvelous improvement in surface quality.

3. Plate Glass

This is the glass used in store windows, mirrors, automobile windows, and aquariums. It is produced in thicknesses of $\frac{1}{8}$ to 2-in. and up to 15 feet wide. The traditional method was to grind and polish the glass on both sides – an expensive process. Today all plate glass, except unusual pieces, is made by the float process (see Chapter 14).

This glass is made with a tinted shade for heat absorption or architectural effects. It also comes in opaque colors for store-fronts and to replace tiles in the bathroom.

4. Container Glass

Bottles and jars come in a variety of shapes and in many colors. The thinner, non-returnable bottles are increasingly used, leading to an ever-worsening litter problem.

Chemical glassware — various sizes of beakers, flasks, graduates, etc. — usually is made of low-alkali glass in order to stand a thermal shock.

5. Tableware and Kitchen Glass

Pressed-glass salad plates, fruit dishes, and dessert plates are made in clear, colored, and opaque glass. A relatively new ware of pressed glass covered with a glaze in compression has an attractive appearance and is practically unbreakable.

In the kitchen borosilicate glass is found in all types of utensil. This ware may be used in the oven but is not reliable for top-of-the-stove use.

6. Art Glass

This ware is found in a great variety of shapes, sizes, and colors. The finer pieces are hand-blown of lead glass, because the high index of refraction gives brilliance and it is softer for cutting and engraving.

7. Special Glasses

Fused silica. This material comes in the form of tubes, flasks, and many other shapes. It may be obtained in a transparent form or the less-expensive opaque. Fused silica is very resistant to temperature shock and can be used in temperatures up to $1200°C$. Above this it starts to devitrify.

Vycor. This glass is melted, formed, and heat-treated to form two phases, one nearly pure silica and other a soluble form that can be leached out. The porous structure is then sintered with considerable shrinkage to give a glass containing about 96% silica. It therefore has good thermal shock resistance and is less expensive than fused silica.

Ultraviolet-transmitting glasses. These are high-silica or phosphate glasses used for solariums.

Solder glasses. There is a whole family of low-fusing glasses with controlled expansion coefficients used for joining glass to glass or glass to metal. Some of the glasses devitrify on cooling, leaving crystals that add to the strength.

8. Optical Glass

These glasses require extraordinary care in manufacture as they must hold to very close optical specifications, be free of bubbles, and be quite homogeneous. Various compositions are produced to give the many optical properties needed.

A new and exacting need for an unusual optical glass is in high-powered lasers. Here the glass must be highly transparent, so only a few parts in a million of impurities are allowed. The rare earth in the glass must not be undistributed by stria and there must not be any inclusions.

9. Fiber Glass

Textile fiber. Large quantities of continuous fibers are spun into yarn and woven into cloth for hangings, fireproof curtains, and furniture coverings. The yarn is also used for reinforcement in plastic and tires. The diameter of a single fiber is about 0.0003 in. and up to 408 strands are drawn simultaneously from a platinum bushing. The glass composition is low in alkalis (E-glass) to resist moisture.

Insulating fibers. These are generally short fibers produced in bulk, pellets, or blankets. The composition may be a soda-lime glass, rock or slag wool, or refractory silica-alumina compositions. They are used for house, refrigerator, pipe, duct, and stove insulation at low or medium temperatures and for furnace insulation at elevated temperatures.

Optical fibers. Glass fibers and rods have been used to transmit light for a long time, but there was much loss along the way through the surface. It has recently been found that, if a fiber with a high index of refraction is coated with a glass of lower index, then the light entering one end, even at an angle, is repeatedly and totally reflected along the way and emerges complete from the other end (Fig.22.1). Such fibers may be drawn to a diameter of 10 microns and to any length. With regular optical glass as core, they are able to transmit light readily up to 100 feet, but with a core glass of super purity transmission of a mile or more is possible. There is great interest in this fiber for message transmission, since at least 10,000 times as many simultaneous messages can be carried over a single glass fiber than over a single copper wire.

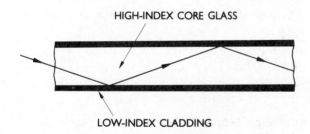

HIGH-INDEX CORE GLASS

LOW-INDEX CLADDING

Fig. 22. 1 Transmission of light through a coated glass fiber.

These coated glass fibers may be put together as a bundle in parallel array, perhaps as many as a million. The ends of the bundle are fused or cemented together and polished. This gives a flexible image transmitter that is used for gastroscopes and many other purposes. The coated fibers may also be welded together to form a vacuum-tight face plate that is used in image intensifiers. A microphotograph of such a face plate is shown in Fig.22.2.

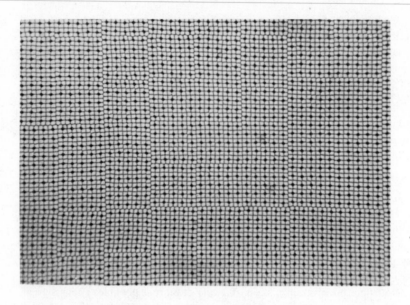

Fig. 22. 2   Microphotograph of a fiber-glass face plate for an image intensifier, 250 X (American Optical Co.).

References

Rauch, H. W., Sr., *et al., Ceramic Fibers and Fibrous Composite Materials*, Academic Press, New York, 1968

Cooper, A. R., and D. A. Krohn, Strengthening of Glass Fibers, II, Ion Exchange, *J. Am. Ceramic Soc.* **52**, 665, 1969

Li, P. C., *et al.*, Fused Quartz Fiber Optics for Ultra Violet Transmission, *Bull. Am. Ceramic Soc.* **48**, 214, 1969

Tummala, R. R., and A. L. Friedberg, Strength of Glass-Crystal Composites, *J. Am. Ceramic Soc.* **52**, 228, 1969

Pearson, A. D., and A. R. Tynes, Light Guidance in Glass Media, *Bull. Am. Ceramic Soc.* **49**, 969, 1970

23

Properties of ceramic materials

1. Introduction

This chapter covers in general the properties of ceramic materials, but there is no intention of giving lists of specific figures for two reasons; first, because available data are not usually accompanied by an exact description of the material tested, and second, because general figures are readily available in the literature. The aim is to give an overall picture of property values and the way they may be altered to fit given conditions.

2. Mechanical Properties

Introduction. Ceramic materials are inherently brittle, so some of the measurements readily made on metals have little meaning in the ceramic field. Although tensile tests can be made on glass fibers, it is difficult to make significant tensile tests on polycrystalline ceramic specimens, so strength tests for the latter are usually made by four-point transverse or compression tests.

It will also be found that repeated tests on a series of identical specimens will show greater individual deviations from the mean than for similar tests with metals. Therefore, to obtain representative values, 10 to 20 identical specimens must be tested.

Transverse strength at room temperature. Triaxial bodies of nonvitreous types have strengths of 2500 to 5000 psi, and vitreous bodies run from 7000 to 10,000 psi. However, when calcined alumina (17%) is added to the vitreous triaxial body, values up to 35,000 psi have been reported. Alumina increases the weight of the body and decreases its translucency, but it is regularly used in hotel china and high-tension electrical insulators. There seems to be no logical explanation for this great increase in strength.

Bone-china bodies have strengths of about the same values as the vitreous triaxial bodies. Steatite, when vitrified, has strengths from 15,000 to 20,000 psi.

Sintered oxides have high strengths when there is little or no glass present. Alumina has values of 40,000 to 100,000, the higher values for bodies with a fine grain structure of 1 to 2 microns. Other pure oxides have values about half that of alumina.

Compared to metals, most glasses are brittle at room temperature and below. Glass has a considerable compressive but a low tensile strength. The theoretical strength of glass may be computed by the bond strengths, and for soda-lime-silicate glass it amounts to over a million psi, a value many times the actual figure. The low tensile strength is caused by Griffith flaws in the surface.

Strength, as well as most other properties in glass, depends on the thermal history of the particular sample studied. The well-documented relation between tensile strength and diameter of fused silica fibers shown in Fig.23.1 is an example. Thin fibers are quenched more rapidly than those with a larger diameter. The more rapidly a fiber is quenched, the higher is its "fictive temperature" and the more it resembles the uniform structure of the melt. Annealing reduces the fictive temperature and results in localized submicroscopic flaws due to local differences in the compaction of the network.

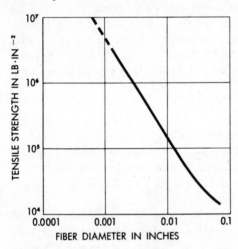

Fig. 23. 1   Breaking strength of silica glass fibers.

The composition of a glass influences its mechanical strength to some extent. The high-softening, strongly bonded glasses like fused silica have the greatest strength, and this decreases as the bonding is weakened.

Glass in the form of sheets, tableware, or ophthalmic lenses has a transverse strength in the annealed condition of 10,000 to 15,000 psi. However, if the surface is given a compression layer by quenching from just below the softening point, the strength can be doubled and when fracture does occur, there are no sharp fragments.

Another method of forming a compression layer, employed on a Na-Al-Si-O glass nucleated with TiO_2 to give a body of nephelite and anatase, is to exchange

Na^+ ions for the larger K^+ ions in the surface. This produces kalsilite, which has a larger specific volume than nephelite, and increases the strength from the original 15,000 psi to 200,000 psi.

A simple way to increase the strength of a glass or a devitrified body is to apply a low-expansion glaze, producing a compression at room temperature.

It should be kept in mind these high strength values are reduced when the surface layer becomes abraded in use.

Modulus of elasticity. Another property of importance is the static modulus of elasticity, which is the stiffness under load. Values for some ceramic materials are:

Boron	90×10^6 psi
Graphite	75×10^6
Alumina	50×10^6
Beryllia	40×10^6
Magnesia	35×10^6
Thoria	25×10^6
Zircon	25×10^6
Triaxial body	20×10^6
Fused quartz	11×10^6
Glass	9×10^6

Impact strengths. This is an important property for ceramics. It is measured by a swinging pendulum or a dropping ball. The values, expressed in foot pounds, depend somewhat on the configuration and the support of the specimen, so no specific values will be given here. However they do relate closely to the transverse strength.

Hardness. The hardness at room temperature varies widely for ceramic materials, as shown below:

Diamond	7000
Silicon carbide	2500
Fused alumina	2000
Beryllium oxide	1250
Zirconium oxide	1100
Quartz	820
Porcelain glaze	600
Sheet glass	530

Particle size. The measurement and control of particle size is extremely important in the field of ceramics. Large and medium-sized particles may be readily measured by using a standard series of screens, as shown in Table 6.1. The smaller particles can be measured by sedimentation methods based on Stokes' law. There are now available rapid automatic methods, such as the Coulter counter, that can work from

0.5 to 400 microns and give a complete distribution curve in less than a minute. Smaller sizes are measured by sedimentation in a centrifuge or by the electron microscope.

Porosity. The total volume of closed pores may be measured by water immersion as described in Chapter 10, Section 7.

The size distribution of the open pores may be determined by immersing the specimen in mercury under varying pressures; the higher the pressure, the smaller the pores filled. Equipment is available to measure pore diameters from 1 mm down to 30 Å, using pressures up to 60,000 psi.

The total porosity, including closed pores, may be calculated from the true and bulk density as shown in Chapter 10, Section 7.

3. Thermal Properties

Hot strength. The strength and rigidity of nearly all materials fall off with increasing temperature. This fact is important when designing high-temperature structures such as furnaces and kilns. Usually, however, rupture strength is not a criterion, as the material flows or creeps long before failure. It is necessary for the designer to know the creep rate at the working temperature in order to be sure the structure will not deform unduly during its normal life. Figure 23.2 shows the creep rate of a refractory at various loads and constant temperature, which plotted on log-log paper gives a straight line. Ceramic materials free from a glassy phase are generally the most resistant to high-temperature deformation.

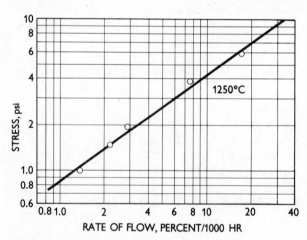

Fig. 23. 2   Flow rate of a refractory versus load.

Coefficient of thermal expansion. The coefficient of linear thermal expansion is the ratio of the change in length per degree C to the length at 0°C. If the length of a solid is plotted against temperature the slope of the resulting line gives the linear coefficient. Volume coefficients, for solids, are close to three times the linear value. It should be understood that we refer here only to *reversible* length changes. In Table 23.1 are given values for the more important pure oxides.

Table 23.1 Expansion coefficients for pure, sintered refractories

Oxide	Expansion coefficient, $^{\circ}$C, 0–1000°C
Alumina	8.6×10^6
Beryllia	8.9×10^6
Boron carbide	4.5×10^6
Magnesia	13.5×10^6
Mullite	5.3×10^6
Silicon carbide	4.7×10^6
Spinel	7.6×10^6
Thoria	9.2×10^6
Titanium carbide	7.4×10^6
Zircon	4.2×10^6
Zirconia	10.0×10^6

In many cases the expansivity is a very important property; for glazes and enamels it must be such as to cause a moderate compression on cooling, and for bodies and glasses a low value is needed to withstand thermal shock.

In the case of glasses, values run from $0.5 \times 10^{-6}\,^{\circ}$C for fused quartz and 3 to $4 \times 10^{-6}\,^{\circ}$C for borosilicates, and up to $9 \times 10^{-6}\,^{\circ}$C for high-alkaline glasses.

Low-expansion bodies have values of $4 \times 10^{-6}\,^{\circ}$C for ovenware and $2 \times 10^{-6}\,^{\circ}$C for top-of-the-stove ware, so it is quite difficult to find glazes to fit them.

Thermal shock resistance. This is an important property of refractories in order that they may stand rapid temperature changes without cracking. For a homogeneous material the tendency for thermal cracking to occur is proportional to:

$$\frac{\alpha}{h\,\theta_b},$$

where α is the coefficient of expansion, h is the thermal diffusivity of the body, and θ_b is the flexibility of the body. In addition to properties of the body itself, the size and shape of the piece will also come into play.

If the body is not homogeneous, such as a grog-clay mixture, there can be set up microstresses due to expansion differences between the different components.

In general, materials such as graphite, some of the carbides, and oxides with a flexible structure are resistant to thermal shock.

Fusion point. This value varies over a large range, but it is of little value except to screen out materials having fusion points below the desired use temperature. Many materials are limited for high-temperature used by volatilization, deformation, and oxidation.

Thermal conductivity. This property (K value) is a difficult one to measure, especially at high temperatures, which accounts for the large scatter of results in the literature.

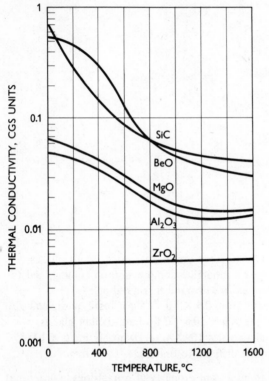

Fig. 23. 3 Thermal conductivity of some pure dense refractories.

In Fig. 23.3 are shown values of some pure, dense ceramic materials. Beryllia is unique among oxides because of its high thermal conductivity. When these materials have porosity, the K values decrease quite rapidly, as shown by the following equation from Loeb (1954):

$$\frac{K_p}{K} = 1 - P_c \frac{1 - \dfrac{4\gamma\,e\,\sigma\,d\,\bar{T}^3}{k}}{1 + \dfrac{4\gamma\,e\,\sigma\,d\,\bar{T}^3}{k} \cdot \dfrac{1 - P_2}{P_2}},$$

where

P_c = fraction occupied by pores of a cross-sectional area cut perpendicularly to the direction of heat flow in a plane containing the pores

P_2 = fraction occupied by pores along the length of a line of heat flow that passes through the pores

γ = a geometric factor depending on shape and orientation of the pores

e = emissivity of pore walls

σ = Stefan's radiation constant

K = conductivity of solid

K_p = conductivity of porous sample

d = dimension of a pore in direction of heat flow

$\bar{T}^3$ = mean cube of absolute temperature of sample.

The estimated types of heat flow through a porous solid are shown in Table 23.2 and the thermal conductivity of some insulators in Table 23.3.

Table 23.2 Heat transfer through a porous solid

Type of transfer	Amount, %
Total	100
By the solid	73
By the gas in the pores	11
By convection in the pores	1
By radiation across the pores	15

Table 23.2 Thermal conductivity,$^*K$, of insulators for medium and high temperatures

Type	K at 500°C	K at 1000°C	K at 1500°C	Upper use temp., °C
Insulating firebrick	0.0011	0.0012	0.0014	1650
Insulating firebrick	0.0008	0.0010	–	1425
Insulating firebrick	0.0003	0.0007	–	1100
Kaolin wool blanket	0.0004	0.0008	–	1100
Rock wool blanket	0.0004	–	–	900
Diatomaceous earth blocks	0.0003	–	–	850
Magnesia asbestos blocks	0.0002	–	–	700

* CGS units

4. Electrical Properties

Introduction. At the start of the century, when only direct current and low frequencies were employed, porcelain was quite satisfactory for insulators, but as interest developed in the handling of high-frequency circuits for radio transmission more efficient materials were sought. This led to the use of low-loss dielectrics like steatite and later insulators with high dielectric constants and ceramics with magnetic properties. Then the internal-combustion engine required spark-plug cores, which encouraged the development and fabrication of new ceramic materials.

Volume resistivity. This property should be high for electrical insulators and low for resistance and heating elements. Fortunately, ceramic materials cover a large range so a selection can usually give the desired value. Most ceramic materials at room temperature are good insulators with volume resistivity values of 10^{12} to 10^{15} ohms-cm, but these values fall off slowly at elevated temperatures. Materials such as tin oxide, molybdenum disilicide, silicon carbide, carbon, and graphite do conduct electrically at room temperature and may be used as resistance and heating elements. The best electrical insulator at high temperatures is pure magnesium oxide.

Dielectric strength. This property is high for the majority of ceramic materials, running between 100 and 200 volts per mil (by ASTM Test no. D667-427), and so most of them are applicable to most high-voltage uses.

Dielectric constant. This property (k_1) is measured by ASTM Test no. D667-42T and represents the ratio of the capacity of a given condenser with the specific material between the plates to that of the same condenser with a vacuum between the plates. The majority of ceramic materials have values between 5 and 8, but it has been known for a long time that TiO_2 crystals showed exceptional values up to 100. During World War II it was found that alkaline earth titanates had extraordinary values of k_1, sometimes up to 10,000. This made possible the production of very small condensers and led to the miniaturization of electronic equipment by printed circuits.

Loss factor. This factor (K^{11}) is represented by the energy loss in ergs per cubic centimeter per cycle. Triaxial bodies have high losses and are unsuitable for use at radio frequencies. As the losses are due largely to the alkaline ions, bodies such as steatites, alumina, and titanates with low-loss factors are now in general use.

Magnetic properties. Ceramic ferromagnetic materials may be classified into "soft" types of the ferrite spinels and "hard" types such as the magnetoplumbite structure. The former are used in transformer cores, where their low loss makes them useful at high frequencies and "square loop" ferrites are used in memory cores for computers. The magnetoplumbites are used as permanent magnets.

5. Optical Properties

These properties are of particular interest in the field of glass. The lens designer has made great progress in the last twenty-five years for two reasons: (1) a host of new glasses have been produced, often with rare-earth components and (2) the computer has enormously speeded up the lens calculation.

Refractive index. This is the ratio of the velocity of light in a vacuum to the velocity in the glass. It can be measured by instruments based on Snell's Law. The principle is shown in Fig.23.4, where

$$N_2 \, S \, \text{in} \, A_2 = N_1 \, S \, \text{in} \, A_1$$
$$\text{(Vacuum)} \qquad \text{(Glass)}$$

$$\text{if } N_2 = 1 \qquad \text{and} \qquad N_1 = \frac{\text{Sin} \, A_2}{\text{Sin} \, A_1} = \text{index of refraction.}$$

Values of N_1 run from 1.35 for very light crown glasses up to 2.10 for very heavy flints. Optical glasses have N_1 controlled within $\pm 3 \times 10^5$.

Dispersion. This property is the spreading of white light into various wavelengths as shown in Fig.23.4(b). Dispersion is given by the Abbe number V_d:

$$V_d = \frac{n_d - 1}{n_f - n_c} \, ,$$

where d is the yellow helium line (587.56 microns), f is the blue hydrogen line (486.13 microns), and c is the red hydrogen line (656.27 microns).
   This property is controlled in optical glass to$\mp 2 \times 10^{-5}$.

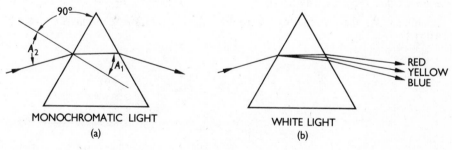

MONOCHROMATIC LIGHT   WHITE LIGHT
(a)   (b)

Fig. 23. 4   Refraction and dispersion of light.

Transmittance. The most useful property of glass is its transparency, which may, however, be impaired by absorption due to transition or rare-earth ions and scattering from small inclusions. Glasses are now being made with only a few parts per million of impurities and they give transmission of light over distances of many meters.

References

Loeb, A. L., A Theory of Thermal Conductivity of Porous Materials, *J. Am. Ceramic Soc.* **37**, 96, 1954

Powell, R. W., Thermal Conductivities of Solid Materials at High Temperatures, *Research* **7**, 492, 1954

Morey, G. W., *The Properties of Glass*, Reinhold, New York, 1959

Kingery, W. D., *Property Measurements at High Temperatures*, Wiley, New York, 1959

Dinsdale, A., *et al.*, The Mechanical Strength of Ceramic Tableware, *Trans. Brit. Ceramic Soc.* **66**, 367, 1967

Davidge, R. W., and G. Tappin, Thermal Shock and Fracture in Ceramics. *Trans. Brit. Ceramic Soc.* **66**, 405, 1967

Storms, E. K., *Refractory Carbides*, Academic Press, New York, 1967

Evans, P. E., *et al.*, Alumina-based Cutting Tools, *Trans. Brit. Ceramic Soc.* **66**, 523, 1967

Alper, A. M., *High Temperature Oxides: Part 1, Magnesia, Lime and Chrome Refractories; Part 2, Oxides of Rare Earth, Titanium and Zirconium; Part 3, Magnesia, Alumina and Beryllia Ceramics*, Academic Press, New York, 1970

Wittemore, O. J., Jr. (ed.), *Electronic Ceramics*, Special Publication no.3, American Ceramic Society, Columbus, Ohio, 1970

Burgman, J. A., and E. M. Hunia, The Effect of Fiber Diameter, Environmental Moisture and Cooling Time During Fiber Formation on the Strength of E Glass Fibers, *Glass Tech.* **11**, 147, 1970

Shahinian, P., High Strength of Sapphire Filament, *J. Am. Ceramic Soc.* **54**, 67, 1971

Appendix

Table A.1 Properties of the atom

Element	Atomic no.	Atomic wt.	Charge	Radius in A	Symbol
Aluminum	13	26.98	+3	0.51	Al
Antimony	51	121.76	−3	2.45	Sb
			+3	0.76	
			+5	0.62	
Arsenic	33	74.92	−3	2.22	As
			+3	0.58	
			+5	0.46	
Barium	56	137.36	+1	1.53	Ba
			+2	1.34	
Beryllium	4	9.01	+1	0.44	Be
			+2	0.35	
Bismuth	83	208.99	+1	0.98	Bi
			+3	0.96	
			+5	0.74	
Boron	5	10.82	+1	0.35	B
			+3	0.23	
Bromine	35	79.92	−1	1.96	Br
			+5	0.47	
			+7	0.39	
Cadmium	48	112.41	+1	1.14	Cd
			+2	0.97	
Calcium	20	40.08	+1	1.18	Ca
			+2	0.99	
Carbon	6	12.01	−4	2.60	C
			+4	0.16	
Cerium	58	140.13	+1	1.27	Ce
			+3	1.03	
			+4	0.92	
Cesium	55	132.91	+1	1.67	Cs
Chlorine	17	35.46	−1	1.81	Cl
			+5	0.34	
			+7	0.27	
Chromium	24	52.01	+1	0.81	Cr
			+2	0.89	
			+3	0.63	
			+6	0.52	
Cobalt	27	58.94	+2	0.72	Co
			+3	0.63	
Copper	29	63.54	+1	0.96	Cu
			+2	0.72	
Dysprosium	66	162.51	+3	0.91	Dy
Erbium	68	167.27	+3	0.88	Er
Europium	63	152.0	+2	1.09	Eu
			+3	0.95	
Fluorine	9	19.00	−1	1.33	F
			+7	0.08	
Gadolinium	64	157.26	+3	0.94	Gd
Gallium	31	69.72	+1	0.81	Ga
			+3	0.62	

Table A.1 *(continued)*

Germanium	32	72.60	− 4	2.72	Ge
			+ 2	0.73	
			+ 4	0.53	
Gold	79	197.0	+ 1	1.37	Au
			+ 3	0.85	
Haffnium	72	178.50	+ 4	0.78	Hf
Holmium	67	164.94	+ 3	0.89	Ho
Hydrogen	1	1.0	− 1	1.54	H
Indium	49	114.82	+ 3	0.81	In
Iodine	53	126.91	− 1	2.30	I
			+ 5	0.62	
			+ 7	0.50	
Iridium	77	192.2	+ 4	0.68	Ir
Iron	26	55.85	+ 2	0.74	Fe
			+ 3	0.64	
Lanthanum	57	138.92	+ 1	1.39	La
			+ 3	1.02	
Lead	82	207.21	+ 2	1.20	Pb
			+ 4	0.84	
Lithium	3	6.94	+ 1	0.68	Li
Lutetium	71	174.99	+ 3	0.85	Lu
Magnesium	12	24.32	+ 1	0.82	Mg
			+ 2	0.66	
Manganese	25	54.94	+ 2	0.80	Mn
			+ 3	0.66	
			+ 4	0.60	
			+ 7	0.46	
Mercury	80	200.61	+ 1	1.27	Hg
			+ 2	1.10	
Molybdenum	42	95.95	+ 1	0.93	Mo
			+ 4	0.70	
			+ 6	0.62	
Neodymium	60	144.27	+ 3	1.00	Nd
Nickel	28	58.71	+ 2	0.69	Ni
Niobium	41	92.91	+ 1	1.00	Nb
			+ 4	0.74	
			+ 5	0.69	
Nitrogen	7	14.01	− 3	1.71	N
			+ 1	0.25	
			+ 3	0.16	
			+ 5	0.13	
Osmium	76	190.02	+ 4	0.88	Os
			+ 6	0.69	
Oxygen	8	16.00	− 2	1.32	O
			− 1	1.76	
			+ 1	0.22	
			+ 6	0.09	
Palladium	46	106.4	+ 2	0.80	Pd
			+ 4	0.65	
Phosphorus	15	30.98	− 3	0.35	P
			+ 3	1.13	
			+ 5	0.98	
Platinum	78	195.09	+ 2	0.80	Pt
			+ 4	0.65	
Plutonium	94	242.0	+ 3	1.08	Pu
			+ 4	0.93	
Polonium	84	210.0	+ 6	0.67	Po
Potassium	19	39.10	+ 1	1.33	K
Praseodymium	59	140.92	+ 3	1.01	Pr
			+ 4	0.90	

Table A.1 *(concluded).*

Rhenium	75	186.22	+4	0.72	Re
			+7	0.56	
Rhodium	45	102.91	+3	0.68	Rh
Rubidium	37	85.48	+1	1.47	Rb
Ruthenium	44	101.10	+4	0.67	Ru
Samarium	62	150.35	+3	0.96	Sm
Scandium	21	44.96	+3	0.73	Sc
Selenium	34	78.96	−2	1.9	Se
			−1	2.32	
			+1	0.66	
			+4	0.50	
			+6	0.42	
Silicon	14	28.09	−4	2.71	Si
			−1	3.85	
			+1	0.65	
			+4	0.42	
Silver	47	107.87	+1	1.26	Ag
			+2	0.89	
Sodium	11	22.99	+1	0.97	Na
Strontium	38	87.83	+2	1.12	Sr
Sulfur	16	32.07	−2	1.84	S
			+2	2.19	
			+4	0.37	
			+6	0.30	
Tantalum	73	180.95	+5	0.68	Ta
Technetium	43	99.0	+7	0.98	Tc
Tellurium		127.61	−2	2.11	Te
			−1	2.50	
			+1	0.82	
			+4	0.70	
			+6	0.56	
Terbium	65	158.93	+3	0.92	Tb
			+4	0.84	
Thallium	81	204.39	+1	1.47	Tl
			+3	0.95	
Thorium	90	232.0	+4	1.02	Th
Tin	50	118.70	−4	2.94	Sn
			−1	3.70	
			+2	0.93	
			+4	0.71	
Titanium	22	47.90	+1	0.96	Ti
			+2	0.94	
			+3	0.76	
			+4	0.68	
Tungsten (wolfram)	74	183.86	+4	0.70	W
			+6	0.62	
Uranium	92	238.07	+4	0.97	U
			+6	0.80	
Vanadium	23	50.95	+2	0.88	V
			+3	0.74	
			+4	0.63	
			+5	0.59	
Ytterbium	70	173.04	+2	0.93	Yb
			+3	0.86	
Yttrium	39	88.91	+3	0.89	Y
Zinc	30	65.38	+1	0.88	Zn
			+2	0.74	
Zirconium	40	91.22	+1	1.09	Zr
			+4	0.79	

Table A.2 Equivalent weights of common ceramic materials

Material	Formula	Formula weight	$\frac{RO}{R_2O}$	R_2O_3	RO_2
			Equivalent weight		
Alumina	Al_2O_3	101.9		101.9	
Ammonium carbonate	$(NH_4)_2 \cdot CO_3 \cdot H_2O$	114.1	114.1		
Antimony oxide	Sb_2O_3	291.5		291.5	
Arsenious oxide	As_2O_3	197.8		197.8	
Barium carbonate	$BaCO_3$	197.4	197.4		
Boracic acid	$B_2O_3 \cdot 3H_2O$	123.7		123.7	
Boric oxide	B_2O_3	69.6		69.6	
Borax	$Na_2B_4O_7 \cdot 10\,H_2O$	381.4	381.4	190.7	
Calcium carbonate (whiting)	$CaCO_3$	100.1	100.1		
Calcium fluoride	CaF_2	78.1	78.1		
Chromic oxide	Cr_2O_3	152.0	76.0	152.0	
Clay (kaolinite)	$Al_2O_3 \cdot 2SiO_2 \cdot 2H_2O$	258.2		258.2	129.1
Cobaltic oxide	Co_2O_3	165.9	83.0	165.9	
Cryolite	Na_3AlF_6	210.0	140.0	420.0	
Copper oxide (cupric)	CuO	79.6	79.6		
Feldspar (potash)	$K_2O \cdot Al_2O_3 \cdot 6SiO_2$	556.8	556.8	556.8	92.9
Feldspar (soda)	$Na_2O \cdot Al_2O_3 \cdot 6SiO_2$	524.5	524.5	524.5	87.6
Flint (quartz)	SiO_2	60.1			60.1
Iron oxide (ferrous)	FeO	71.8	71.8		
Iron oxide (ferric)	Fe_2O_3	159.7	79.8	159.7	
Lead carbonate (white lead)	$2PbCO_3 \cdot Pb(OH)_2$	775.6	258.5		
Lead oxide (red lead)	Pb_3O_4	685.6	228.5		
Lithium carbonate	Li_2CO_3	73.9	73.9		
Magnesium carbonate	$MgCO_3$	84.3	84.3		
Magnesium oxide	MgO	40.3	40.3		
Manganese dioxide	MnO_2	86.9	86.9		86.9
Nickel oxide	NiO	74.7	74.7		
Potassium carbonate	K_2CO_3	138.0	138.0		
Sodium carbonate	Na_2CO_3	106.0	106.0		
Strontium carbonate	$SrCO_3$	147.6	147.6		
Tin oxide	SnO_2	150.7			150.7
Titanium dioxide	TiO_2	79.9			79.9
Zinc carbonate	$ZnCO_3$	125.4	125.4		

Table A.2 *(concluded)*

Zinc oxide	ZnO	81.4	81.4	
Zirconium oxide	ZrO_2	123.2		123.0

Table A.3 Temperature-conversion table *

°C	0	10	20	30	40	50	60	70	80	90		
	F	F	F	F	F	F	F	F	F	F		
−200	−328	−346	−364	−382	−400	−418	−436	−454	.....	.....		
−100	−148	−166	−184	−202	−220	−238	−256	−274	−292	−310		
− 0	+ 32	+ 14	− 4	− 22	− 40	− 58	− 76	− 94	−112	−130		
0	32	50	68	86	104	122	140	158	176	194	°C	°F
100	212	230	248	266	284	302	320	338	356	374	1	1.8
200	392	410	428	446	464	482	500	518	536	554	2	3.6
300	572	590	608	626	644	662	680	698	716	734	3	5.4
400	752	770	788	806	824	842	860	878	896	914	4	7.2
500	932	950	968	986	1004	1022	1040	1058	1076	1094	5	9.0
600	1112	1130	1148	1166	1184	1202	1220	1238	1256	1274	6	10.8
700	1292	1310	1328	1346	1364	1382	1400	1418	1436	1454	7	12.6
800	1472	1490	1508	1526	1544	1562	1580	1598	1616	1634	8	14.4
900	1652	1670	1688	1706	1724	1742	1760	1778	1796	1814	9	16.2
1000	1832	1850	1868	1886	1904	1922	1940	1958	1976	1994	10	18.0
1100	2012	2030	2048	2066	2084	2102	2120	2138	2156	2174		
1200	2192	2210	2228	2246	2264	2282	2300	2318	2336	2354		
1300	2372	2390	2408	2426	2444	2462	2480	2498	2516	2534	°F	°C
1400	2552	2570	2588	2606	2624	2642	2660	2678	2696	2714	1	0.56
1500	2732	2750	2768	2786	2804	2822	2840	2858	2876	2894	2	1.11
1600	2912	2930	2948	2966	2984	3002	3020	3038	3056	3074	3	1.67
1700	3092	3110	3128	3146	3164	3182	3200	3218	3236	3254	4	2.22
1800	3272	3290	3308	3326	3344	3362	3380	3398	3416	3434	5	2.78
1900	3452	3470	3488	3506	3524	3542	3560	3578	3596	3614	6	3.33
2000	3632	3650	3668	3686	3704	3722	3740	3758	3776	3794	7	3.89
2100	3812	3830	3848	3866	3884	3902	3920	3938	3956	3974	8	4.44
2200	3992	4010	4028	4046	4064	4082	4100	4118	4136	4154	9	5.00
2300	4172	4190	4208	4226	4244	4262	4280	4298	4316	4334	10	5.56
2400	4352	4370	4388	4406	4424	4442	4460	4478	4496	4514	11	6.11
2500	4532	4550	4568	4586	4604	4622	4640	4658	4676	4694	12	6.67
2600	4712	4730	4748	4766	4784	4802	4820	4838	4856	4874	13	7.22
2700	4892	4910	4928	4946	4964	4982	5000	5018	5036	5054	14	7.78
2800	5072	5090	5108	5126	5144	5162	5180	5198	5216	5234	15	8.33
2900	5252	5270	5288	5306	5324	5342	5360	5378	5396	5414	16	8.89
3000	5432	5450	5468	5486	5504	5522	5540	5558	5576	5594	17	9.44
3100	5612	5630	5648	5666	5684	5702	5720	5738	5756	5774	18	10.00
3200	5792	5810	5828	5846	5864	5882	5900	5918	5936	5954		
3300	5972	5990	6008	6026	6044	6062	6080	6098	6116	6134		
3400	6152	6170	6188	6206	6224	6242	6260	6278	6296	6314		
3500	6332	6350	6368	6386	6404	6422	6440	6458	6476	6494		
3600	6512	6530	6548	6566	6584	6602	6620	6638	6656	6674		
3700	6692	6710	6728	6646	6764	6782	6800	6818	6836	6854		
3800	6872	6890	6908	6926	6944	6962	6980	6998	7016	7034		
3900	7052	7070	7088	7106	7124	7142	7160	7178	7196	7214		

Examples. 1347°C = 2444°F + 12.6°F = 2456.6°F ; 3367°F = 1850°C + 2.78°C = 1852.78°C.

*From Dr. L. Waldo, *Metallurgical and Chemical Engineering,* March 1910.

Reproduced by permission from *Refractories,* 3rd ed., by F. H. Norton. Copyright, 1949, McGraw-Hill Book Company, Inc.

Table A.4 Temperature equivalents of Orton pyrometric cones
(large cones, 150°C/hr)*

Cone no.	Endpoint, $^\circ$C	Cone no.	Endpoint, $^\circ$C
022	600	10	1305
021	614	11	1315
020	635	12	1326
019	683	13	1346
018	717	14	1366
017	747	15	1431
016	792	16	1473
015	804	17	1485
014	838	18	1506
013	852	19	1528
012	884	20	1549
011	894	23	1590
010	894	26	1605
09	923	27	1627
08	955	28	1633
07	985	29	1645
06	999	30	1654
05	1046	31	1679
04	1060	32	1717
03	1101	$32\frac{1}{2}$	1730
02	1120	33	1742
01	1137	34	1759
1	1154	35	1784
2	1162	36	1796
3	1168	37†	1830
4	1186	38†	1850
5	1196	39†	1865
6	1222	40†	1885
7	1240	41†	1970
8	1263	42†	2015
9	1280		

*Taken from *J. Am. Ceramic Soc.* **39**, 47, 1956.
†Small cones, 600°C/hr.

Author index

Adcock, D. S., 113
Akroyd, T. N. W., 277
Alietti, A., 23
Allen, A. C., 162
Alper, A. M., 259, 292
Andrews, A. I., 5, 222
Asbury, N. F., 91
Ayer, J. E., 113

Barber, E. A., 5
Bates, T. F., 37
Benbow, J. J., 113
Berard, M. F., 6
Borden, W. J. H., 113
Blackburn, A. R., 113
Blakeley, J. D., 259
Bloor, E. C., 91, 210, 251
Bowles, D., 41
Bragg, W. L., 11, 13
Brett, N. H., 152
Brindley, G. W., 136, 151, 152
Brown, S. D., 182
Bruce, R. H., 54, 152
Buchkemer, H., 162
Buchnaner, R., 162
Budnikoz, P. P., 6
Budworth, D. W. A., 153, 210

Burgman, J. A., 292
Burke, J. E., 6, 152
Bush, E. A., 250
Byrne, J., 113

Cable, M., 183
Calhoun, W. A., 71
Callister, W. D., 152
Carruthers, T. G., 152
Cassidy, D. C., 192
Clark, N. O., 37
Clarke, O. M., 37
Clough, A. R. J., 251
Coble, R. L., 152
Coes, L. Jr., 268
Comer, J. J., 152
Cooper, A. R., 282
Cox, W. E., 5
Cutler, I. B., 113

Dal, P. H., 113
Davidge, R. W., 292
Davis, P., 191
Demiray, T., 238
Deri, M., 251
Dinsdale, A., 152, 292

Dodd, A. E., 6
Dollimore, D., 125
Dubois, H. B., 41

Earhart, W. H., 210
Edward, H., 210
El-Hemaly, S. A., 278
Eppler, R. A., 222, 238
Evans, P. E., 292
Evans, W. D. J., 238
Everett, D. W., 113

Fisher, E. H., 222
Floyd, J. R., 268
Fraas, F., 71
Franklin, C. E. L., 210, 238
Friedberg, 217, 218, 219, 282
Fulrath, R. M., 6, 152

German, W. L., 151, 277
Gibbon, D. L., 216
Gibbs, G. V., 128, 152
Gill, R. M., 113
Gitter, A. J., 71
Gjostein, N. A., 192
Glang, R., 251
Goodeve, C. F., 91
Goodman, G., 151
Goodwin, J. W., 91
Gould, R. E., 250
Green, C. H., 182
Gregg, S. J., 125
Griffiths, R., 5
Grim, R. E., 6, 23, 37
Grindey, W. T., 238
Gruner, J. W., 54
Guillat, I. F., 152

Hainback, R., 237, 238
Hair, M., 172
Hall, C. E., 6, 121
Hall, D., 277
Hamilton, P. K., 23

Hana, K. M., 278
Hardy, A. C., 238
Hargreaves, E. M., 268
Hart, P. E., 251
Harvon, M. A., 183
Hauth, W. E. Jr., 112
Heath, D. L., 251
Heffelfinger, R. E., 54
Helzel, M., 182
Hendricks, S. B., 15, 22
Herrman, E. R., 113
Heystek, H., 54
Hinckly, D. N., 23
Holden, A. N., 259
Holdridge, D. A., 125, 153, 277
Holscher, H. H., 194
Houseman, J. E., 278
Huggett, L. G., 259
Hulbert, S.F., 251
Hummel, F. A., 250
Hunia, E. M., 292

Johns, W. D., 137, 152
Jorgensen, P. J., 251
Johnson, H. B., 152
Jones, J. T., 6
Joyce, I. H., 91
Justice, C. W., 41

Kannoff, A. L., 259
Keeling, P. G., 23, 37, 41
Kerr, P. F., 22, 23
Kessler, F., 152
King, B. W. Jr., 192
Kingery, W. D., 6, 292
Koenig, C. J., 278
Koenig, J. H., 210
Koltermann, M., 152
Kreidel, N., 172
Krohn, D. A., 282
Kruger, O. L., 259

Ladoo, R. B., 5
Lancaster, B. W., 125

Landorf, J. H., 162
Lea, F. M., 278
Lepie, M. P., 210
Levine, E. M., 6
Li, P. C., 282
Lillie, H. R., 170
Lissenden, A., 71
Lundlin, S. T., 251
Loeb, A. L., 292

McCartney, E. R., 125
McDowall, I. C., 113
McFadden, C. A., 102
McKee, R. I., 268

Macey, H. H., 125
Mackenzie, J. D., 172
Maissel, L. I., 251
Manning, J. H., 6
Marboe, C. E., 6, 172
Marks, L. S., 77, 91
Mellor, J. W., 209
Midgley, H. G., 277
Miller, W. J., 101, 102
Moore, F., 125
Morey, G. W., 163, 292
Mozzi, R. L., 172
Mudd, S. W., 5
Mulroy, B., 238, 251
Myers, W. M., 5

Nakahira, M., 136, 151
Newham, R. E., 23
Nicholson, P. S., 153
Norman, J. E., 41
Norris, A. W., 210
Norton, F. H., 6, 37, 210, 237, 238, 251, 259

Omeara, R. G., 41
Onada, G. Y. Jr., 182
Orchard, D. F., 277
Ougland, R. M., 152
Ovenston, A., 113

Page, B. N., 54
Parmelee, C.W., 210
Pask, J. A., 6, 54, 125, 152
Peace, N., 238
Petri, F., 162
Pearson, A. D., 282
Peray, K., 278
Phillips, F. C., 22
Pincus, A., 166
Plann, E., 54
Platts, D. R., 182
Poch, W., 152
Popper, P., 259
Powell, H. E., 71
Powell, R. W., 292
Powers, W. H., 278
Prabhakaram, P., 152
Promisel, N. E., 6

Radford, C., 5, 54
Rado, P., 238
Ratcliffe, S. W., 277
Rauch, H. W. Sr., 282
Reed, R. J., 162
Reeves, J. E., 41
Reingen, W., 113
Reis, H., 5
Reising, J. A., 210
Remmey, G. B., 162
Rhodes, J. R., 192
Riebling, E. F., 182
Rigby, G. R., 22
Rockwell, P. M., 71
Ross, C. S., 22
Ross, W. A., 15, 153
Rouch, H. W. Sr., 6
Roy, R., 152
Ruddlesden, S. N., 152
Russow, J., 23
Rutten, M. G., 54
Ryan, W., 5, 91, 125

St. Pierre, 151
Schairer, J. F., 41
Schmidt, W. G., 251

Scott, B., 152
Scott Blair, G. W., 78, 91
Searle, A. B., 238
Seger, H. A., 194, 239
Seider, R., 251
Sevratosa, J. M., 23
Shahinian, P., 292
Shaw, K., 6, 238, 278
Sheraf, D., 35
Sherwood, T. K., 117, 125
Singer, F., 6
Singer, S. S., 6
Smith, J. W., 216
Solon, L. M., 5
Soppet, F. E., 113
Sortwell, H. H., 37, 210
Sosman, R. B., 54, 152
Speil, S., 37
Stanforth, J. E., 183
Stewart, G. H., 6
Stiglich, J. J., 251
Storms, E. K., 292
Stull, R. T., 210
Sun, K., 164, 172
Svec, J. J., 277

Taeler, D. H., 259
Tamman, G., 171
Tappin, G., 292
Tatnall, R. F., 162
Thurnauer, H., 113
Tindall, J. A., 238
Tischen, R. E., 172
Tokar, M., 251

Tynes, A. R., 282

Van Schoik, E. C., 6
Van Vlack, L. H., 6
van Wunik, J., 113
Volf, M. B., 6, 172
Von Tschirnhaus, G., 4
Voss, W. C., 277

Waddell, J. J., 278
Walker, E. G., 153, 277
Wasner, M. F., 54
Warren, B. E., 6, 13, 172
Warshaw, S. I., 251
Watts, A., 195
Weber, J. N., 152
Wehrenberg, T. M., 113
Weiss, A., 23
Weyl, W. A., 6, 172
Weymouth, J. J., 91
Wilkinson, W. T., 54, 152
Wilkinson, W. O., 91, 125
Witt, J. C., 277
Wittemore, O. J. Jr., 6, 292
Woo, D., 125
Worral, W. E., 33, 37, 91

Yount, J. G. Jr., 278

Zachariasn, W. H., 164, 172

Subject index

Abrasives
  coated, 268
  loose, 268
  natural, 260
Adams and Williamson Law, 170
Adherance, theory of, 212
Albany slip clay, 247
Albite, 38
Alumina
  firing methods for, 144, 145
  for refractories, 255
  fused, 262
  properties of, 51
  use of, 50
Aluminous minerals, hydrated, 20
Aluminum Company of America, 51
American National Standard Institute,
  265
American Optical Corporation, 282
Andalusite, 50, 138
Anhydride, 275
Annealing, 191
Anorthite, 38
Application of glazes, 199, 200, 201
Armor, ceramic, 247

Artware
  body for, 245
  setting of, 149, 151
Atomic bonds, 10
Atoms
  coordination number of, 11, 12
  properties of, 8
  radius ratio for, 12
  structure of, 9
Auger, 104
Avco Corporation, 145, 146

Babcock and Wilcox Company, 148,
  259
Ball clay, deposits of, 26
Barium titanate, 244
Base exchange capacity, 29, 30, 31, 32
Basic brick, 143, 253
Belleek, 245
Beryllia
  for refractories, 255
  uses for, 50
Bickley Furnaces, Inc., 155, 156
Binary glasses, 167

Bioceramics, 243
Blowing Glass, 188
Bodies
    composite, 243
    compositions of, 239
    electrical, 243
    electrical conducting, 244
    fine ceramic, 93
    heavy clay products, 93
    low expansion, 245
    preparation of, 92
    preparation flow sheet, 93
    refractory, 93
    stoneware, 247
    triaxial, 241
Bonds, 10, 256
Bone ash, 41
Bone china
    firing properties of, 142
    nature of, 245
Boric oxide glass, 166
Boron carbide, 256
Brick
    basic, 253
    fire-clay, 252
    high Alumina, 252
    sand-lime, 271
    silica, 252
Brownian movement, 75
Burnishing, 112
Carbides, 256
Carbon, 257
Carbonates, thermal properties of, 139
Carborundum Company, 266, 267, 268
Casting, mechanism of, 99
Casting slips, properties of, 95
Casting under pressure, 98, 99
Cement
    high alumina, 275
    mechanism of setting of, 274
    oxychloride, 277
    phosphate, 277
    Portland, 272
    silicate, 277
Ceramic industry, magnitude of, 4, 5
Ceramic stains, 235, 236, 237
Cesium chloride, unit cell of, 11

Champlevé, 213
Chemical composition of clay, 33, 34, 35
Chemical treatment of raw materials, 64, 65
China, early ceramics of, 1
Chisholm, Boyd and White Company, 106
Chrome ore, chemical analysis of, 48
Chrome ore, deposits of, 48
Chromite, 47
Classification operations, 63
Classification, water, 64
Classifiers, air, 62
Clay
    accessory minerals of, 32, 33
    chemical composition of, 33, 34, 35
    disintegration of, 64
    impurities in, 137
    organic matter in, 33
    particle size of, 28, 29
    particle shape of, 28
    plastic properties of, 36
    slaking properties of, 36
    workability of, 79
Clay minerals, 15
Clay types
    Albany slip, 247
    ball, 26
    flint fire, 27
    plastic fire, 27
    stoneware, 27
Clinoenstatite, 243
Cloisonné, 213
Coesite, 138
Coefficient of thermal expansion, 286, 287
Color
    chromophore, 231
    colloidal, 233
    crystal, 235
    definition of, 229
    elements for, 231
    measurements of, 229, 230
    modifiers of, 231
    solution, 231, 232
Colloid, 72
Colloidal chemistry, 72

Colloidal colors, 233, 234, 235
Comminuting, equipment for, 58, 59
Comminution, principles of, 57
Composition of glazes, 193
Concrete, refractory, 158
Copenhagen, 4
Coordination number for atoms, 11, 12
Cordierite body, 142
Cornish stone, 40, 41
Corundum, 50, 260
Crawling, 209
Crazing, 205
Cristabolite, 45, 138
Crystal
    growth rate of, 134
    single, 248, 249
    structure of, 12, 13
    unit cell of, 10, 11
Crystals in glaze, 202, 203
Crystallographic systems, 13
Cutting tools, 247
Czochralski method, 248, 249

Day tanks, 179
Deflocculant, 96
Deflocculation
    theory of, 73, 74, 75
    mechanism of, 76
Devitrification, 171, 172
Decal, 225
Decalcomania, 225
Decolorizing glass, 177, 178
Decomposition of clay impurities, 137
Decoration
    banding, 228
    hand, 226
    by photography, 226
    by printing, 225
    relief, 223
    sgraffito, 228
    by stamping, 228
    stencil, 227
Diamond, 52, 53, 261, 263
Diaspore
    deposits of, 27
    mineralogy of, 22, 50

Diatomaceous earth, 258, 261
Diatomite, 46
Dielectric constant, 290
Dielectric drying, 125
Dielectric strength, 290
Dilatency, 82
Dilatometer, 135
Disintegration of clay, 64
Dispersion, optical, 291
Dolomite, 47
Drain casting, 97
Drawing sheet glass, 189
Dry pressing, 105
Dry press bodies, flow sheet of, 94
Drying rate, 116, 120
Drying shrinkage
    control of, 120
    mechanism of, 114, 117, 118
Drying raw materials, 66
Dryer
    humidity, 121
    rotary, 67
    spray, 124, 125
    table ware, 124
Dust pressing, 109

E-glass, 281
Efficiency, kiln, 159, 160
Eggshell porcelain, 111
Elastic state, 78
Electrical porcelain, setting of, 147
Electrochemistry, early, 4
Electrodes for glass melting, 181
Elements, abundance of, 53
Electric furnace, carbide, 263
Electro Refractories, 253
Emery, 260
Equilibrium diagrams
    general, 128, 129, 130, 131
    silica alumina, 131
    solid solution, 131
Enamel
    base metal for, 214, 219
    cast iron, 219
    cover coat, 216, 217, 220
    for aluminum, 222

frits for, 212
ground coat, 214, 219
jewelry, 213, 214
opacifiers, 212, 213
refractory, 222
sheet steel, 214
England, early ceramic developments, 4
England, imports from, 252
Eucryptite, 245
Expansion, 134
Extrusion, piston, 105
Eye, human, 229

Feeders, 184
Feldspar
  abrasives, 261
  chemical composition of, 40
  flow sheet for milling of, 69
  minerals of, 38, 39
  mining of, 39
  occurrence of, 39
  stability of, 139
Ferrites, 244
Fiberglass, 190
Filter press, 67
Filtering, 66
Finishing fired ware, 151
Fining glass, 176, 177
Fire brick, insulating, 254
Fire clays, deposits of, 26, 27
Fired ware, finishing of, 151
Firing shrinkage, porcelain, 141
Flint glass, 173
Float glass, 189
Flow
  laws of, 76, 77
  measurement of, 87
  of flocced suspensions, 83
Flow sheet of
  body preparation, 93
  ceramic stains
  dry pressed bodies, 94
  dry process cast iron enamel, 221
  European washed kaolin, 69
  feldspar, 69
  Georgia kaolin, 68

gypsum plaster, 276
Portland cement, 274
refractory grog, 70
sea water magnesia, 70
sheet steel enamel, 220
talc, 71
Flue linings, 271
Fluorite, 41
Fluxing materials, analysis of, 41
Forehearth, 184
Forces between kaolinite particles, 76
Forming methods, 92
Frit porcelain, 245
Frits, enamel, 212, 215
Fritted glazes, compounding of, 197,
      198, 199
Froth floatation, 66
Fuel element, nuclear, 256

Ganister, deposits of, 45
Garnet, 261
Gibbsite
  deposits of, 27, 50
  structure of, 20
  thermal effects for, 138
  unit cell of, 20
Gilding, 237
Glass
  annealing of, 168, 170
  art, 280
  binary, 167
  boric oxide, 166
  commercial, 173
  composition of, 173, 174
  container, 279
  decolorizing of, 177, 178
  devitrification of, 171, 172
  drawing of, 189
  fining of, 176, 177
  finishing of, 191
  flint, 173
  float process, 189
  flow properties of, 169, 170
  grinding of, 191
  insulating fiber, 281

melting mechanism of, 175, 176
melting range of, 168
network in, 163, 164
optical, 173, 380
optical fiber, 281, 282
phosphate, 167
plate, 279
pressing of, 185
quenched, 171
sheet, 279
single component, 166
solder, 280
softening point of, 168
stirring of, 178
tableware, 280
ternary, 167
textile fiber, 280, 281
thermal expansion of, 167, 168
viscosity of, 168
wool from, 258
working range of, 168
Glass blowing
    Lynch machine, 187
    methods of, 188
    Owens machine, 186
Glass batch, 175, 176
Glass composition, expression of, 166
Glass fiber forming, 190, 191
Glass forming oxides, 164, 165
Glass intermediate formers, 165
Glass molds, 188
Glass network formers, 164
Glass network modifiers, 165
Glass pots, 178, 179
Glass sand
    chemical analysis of, 44
    screen analysis of, 44
    specifications of, 44
Glass tank
    container, 179, 180
    ophthalmic, 182
    sheet, 181
Glaze
    application of, 199
    Bristol, 206
    compositions of, 193, 194, 195
    crystaline, 208

defects in, 209, 210
examples of, 206, 207, 208, 209
fritted, 206
life history of, 201
optical properties of, 196
orange peel on, 210
pin holes in, 209
porcelain, 208
raw lead, 206
reduction, 209
specks in, 210
spraying of, 200
stresses in, 204, 205
surface of, 196, 202
wavy surface of, 210
Graphite
    as a refractory, 257
    pyrolytic, 257
    sources of, 50
    unit cell of, 51
Graphic granite, 40, 41
Grinding wheels
    forming of, 264, 265
    safety of, 265
    types of, 263
Grinding, mechanism of, 266
Grindstone, 261
Grog, flow sheet, 20
Ground laying, 226
Gypsum plaster, 275, 276

Harper Electric Furnace Corporation,
    160
Hardness, 52, 285
Heat balance of kilns, 161
Heat of formation, 126, 127
Heat required to fire ware, 162
Hemihydride, 275
History of ceramics, 1, 2, 3
Homer-Laughlin China Corporation,
    101
Hot floor, 122
Hot pressing, 102
Hot pressing of pure oxides, 145
House of cards structure, 73
Humidity dryer, 121, 122, 123

Hydrocyclone, 64, 65
Hydromet American Inc., 108

Insulating firebrick, 254
Insulating materials, 258
Internal flow in drying, 114

Jigger, automatic, 102
Jiggering process, 100, 101
John Welland and Son, 107
Jointing, 112
Josiah Wedgewood and Son, 246

Kaolin
  deposits of, 25
  flow sheet for washing, 68, 69
  origin of, 24
Kaolin wool, 258
Kaolinite, 15, 16, 17
Kanthol heating element, 155
Keatite, 138
Kiln
  controlled atmosphere tunnel, 159
  efficiency of, 159, 16
  elevator, 154, 155
  heat balance of, 161
  muffle tunnel, 159
  rotary, 157
  scove, 157
  shuttle, 155, 156
  stationary, 154
  tunnel, 157, 158, 159
Kyanite, 50, 138

Lanthia, 255
Lathe, automatic, 112
Laws of flow, 76, 77
Lennox, 254
Lime, 47, 271, 272
Light, properties of, 228, 229
Limoges, 213, 214
Lithia body, firing properties of, 142
Literature, ceramic, 4, 5, 6, 7

Liverpool, 225
Loss factor, 290
Lusters, 237
Lyosphere around kaolinite, 73

Magnesia
  for refractories, 255
  sea water process, 70
Magnesite, 46, 47
Magnetic properties of ceramic
    materials, 290
Magnetic separation of minerals, 66
Magnetoplumbite, 244
Measurement of
  flow, 88
  porosity, 135
Meissen, 4
McGraw-Hill building, 271
Mica
  glass bonded, 245
  thermal properties of, 139
Mineralogical formula, 239, 240
Minerals
  accessory, 32, 33
  clay, 15
  feldspar, 38, 39
  hardness of, 52
  three layer, 18
Mining
  open pit, 55
  operations, 57
  underground, 57
Modulus of elasticity, 285
Moisture distribution in drying, 115,
    116
Molds, plaster, 99
Montmorillonite, structure of, 18
Mortars
  for building brick, 272
  for refractories, 258
Mullite, 255
Munssel color system, 229
Muscovite
  structure of, 18
  unit cell of, 19
Murray printing machine, 225

Nepheline seyenite, 40, 41
Network formers, 164
Nitrides, 257
Nonoxide refractories, 258

Opacifiers
  enamel, 212
  properties of, 213
Organic matter in clay, 33
Optical glass, 174, 175, 280
Optical properties of glass, 175
Orthoclase, 38

Parian porcelain, 243
Particle
  packing of, 89, 90, 91
  shape in suspensions, 83
  size of in clay, 29
  size, measurement of, 285, 286
Paving blocks, 271
Perovskite structure, 22
Phase rule, 128
Phosphate glass, 167
Pilkington Bros., 189
Pin holes, 272, 273
Plaster molds, 99
Plastic mass
  forces in, 85, 86
  properties of, 84, 85
  water films in, 85
Plastic state, 78
Plasticity, mechanism of, 84
Plastics, refractory, 258
Poise, 77
Porcelain, start in Europe, 4
Porosity measurement, 135
Portland cement, 209
Press
  hydraulic, 108
  isostatic, 109, 110
  tile, 107
  toggle, 106
Pressing glass, 188
Pressure casting, 98, 99
Pure oxides, hot pressing of, 145

Pyrophyllite, 49

Quartz
  as an abrasive, 261
  deposits of, 43
  properties of, 42, 43
  thermal stability of, 138
  use of, 45
Quenched glass, 171

Radius ratio of atoms, 12
Ram process of forming, 102, 104
Rate of reaction, 132
Raw glazes, compounding of, 196, 197
Reaction rate, 132
Refractive index, 291
Refractories
  fused, 256
  non oxide, 256
  pure oxide, 255
  setting of, 147, 148
Refractory body, firing properties of,
  143
Refractory shapes, 254
Resistivity, volume, 290
Rheopectsy, 82
Richard Remmey Son, Company, 264
Roller forming, 102
Rolling sheet glass, 189
Rouge, 261
Royal Danish Pottery, 227

Sandwich Glass Company, 185
Sadler and Green, 225
Scheidhaur and Geissing process, 90
Screens, 60, 61
Sculpture, 224, 225
Sea water magnesia, flow sheet of, 70
Seam treatment, 111
Setters, refratory, 253
Setting methods, 145, 146, 147
Sèvres, 239, 240
Sewer pipe, 271

Shrinkage
    conversion of, 119, 120
    curves of clays, 118
    measurement of, 134
Silica
    density of phases in, 139
    fused, 280
    hydrated, 46
    phases of, 138
    refractories of, 255
    uses of, 42
Silica brick
    composition of, 252
    firing properties of, 143
Silicate structures, 13, 14
Silicides, 252
Silicon carbide, 261, 262
Sillimanite, 50, 138
Sintered alumina, 144
Sintered beryllia, 145
Sintered magnesia, 145
Slag wool, 258
Slaking properties of clay, 36
Slip casting, 95
Slips for heavy refractories, 89
Soft mud brick, forming of, 100
Solid solutions, 13
Solid casting, 98
Solid-state reactions, 133, 134
Solid-state sintering, 143, 144
Sorel cement, 277
Spinel, unit cell of, 21, 22
Spinks Clay Company, 35
Spode, 4
Spodumene, 245
Spruce Pine, N. C., 25
Stains, 235, 236
Stishovite, 138
Sticking up, 112
Stockbarger method, 248
Stoneware, 247
Storage of raw materials, 68
Strength
    hot, 286
    impact, 285
    traverse, 283, 284
Stress-strain diagrams for clay, 78

Stresses in a glaze, 204, 205
Structure of silicates, 13, 14
Sulphides, 257
Surface evaporation, 116
Swindell-Dressler Corporation, 150

Tableware, setting of, 147
Talc
    chemical analysis of, 49
    deposits of, 49
    flowsheet for flotation of, 71
    stability of, 139
    structure of, 48
Talc body, firing properties of, 142
Terra-cotta, 271
Thermal analysis, kaolin, 137
Thermal conductivity, 288, 289
Thermal expansion of glass, 167, 168
Thermochemical reactions in kaolin,
        136
Thermodynamics, 126
Thoria, 255
Tile
    drain, 271
    press for, 107
    quarry, 271
    roofing, 271
    setting of, 147
    wall, 245
Tin oxide, 50, 255, 256
Toggle press, 106
Translucency, of triaxial body, 140, 142
Transmittance, optical, 291
Triaxial body
    composition of, 241, 242, 243
    heat effects in, 140
    life history of, 140
    translucency of, 140
Tridymite, 45, 138
Trimming, 110
Turning, 111
Tyler series screens, 61

Unit cell of crystals, 10, 11
Urania, 255

Vapour pressure of oxides, 126, 127
Vermiculite, 258
Verneuil method, 248, 249
Viscosity of casting slips, 96
Viscosity of glass, 168
Viscosity of kaolinite suspensions, 74
Viscosity of water, 115
Viscous state, 77

Water films in plastic masses, 85
Water hull, 72
Water, viscosity of, 115

Waterfall glazing, 200, 201
Wedgewood, 4, 67, 246
Wollastonite, 49
Workability of clay, 79

Yield point of flocced suspensions, 83
Yttria, 255

Zircon for refractories, 255
Zirconia for refractories, 255
Zone melting, 249, 250